sma8003623
11/98
£14.95
RWH G
(Ben)

THE GREENING OF MARXISM

DEMOCRACY AND ECOLOGY
A Guilford Series

*Published in conjunction
with the Center for Political Ecology*

JAMES O'CONNOR
Series Editor

IS CAPITALISM SUSTAINABLE?
POLITICAL ECONOMY AND THE POLITICS OF ECOLOGY
Martin O'Connor, Editor

GREEN PRODUCTION
TOWARD AN ENVIRONMENTAL RATIONALITY
Enrique Leff

MINDING NATURE
THE PHILOSOPHERS OF ECOLOGY
David Macauley, Editor

THE GREENING OF MARXISM
Ted Benton, Editor

THE GREENING
OF MARXISM

Edited by Ted Benton

THE GUILFORD PRESS
New York   London

Printed in the United States of America

This book is printed on acid-free paper.

Last digit is print number: 9 8 7 6 5 4 3 2 1

Library of Congress Cataloging-in-Publication Data

The Greening of Marxism / edited by Ted Benton.
    p. cm.–(Democracy and ecology)
    Includes bibliographical references and index.
    ISBN 1-57230-118-X.–ISBN 1-57230-119-8 (pbk.)
    1. Communism and ecology. 2. Capitalism–
Environmental aspects. I. Benton, Ted. II. Series.
HX550.E25G74   1996
333.7–dc20   96-34197
                                        CIP

Permission to reprint the following chapters and commentaries is gratefully
acknowledged:
    From *Capitalism, Nature, Socialism,* © 1988, 1992, 1993, 1994 by Guilford
Publications: Chapter 2, from Vol. 3(1), 19–35 (translated by John N. K. Esseboe, of
the University of Port Harcourt in Nigeria, and by K. P. Moseley, of the University of
Connecticut); Chapter 3, from Vol. 4(2), 65–84; Chapter 5, from Vol. 4(4), 69–88;
Chapter 6, from Vol. 5(4), 95–104 (translated by Jay Caplan); Chapter 7, from Vol.
4(1), 44–66 (translated by Ruth MacKay); Chapter 9, first published as "Capitalism,
Nature, Socialism: A Theoretical Introduction," from Issue 1, Fall, 1988, 11–38;
Commentary 1, from Vol. 3(3), 84–88; Commentary 2, from Vol. 3(3), 92–94;
Commentary 3, from Vol. 4(1), 102–10; Commentary 4, from Vol. 4(4), 117–121;
Chapter 10, from Vol. 3(2), 43–62; Commentary, first published as a "Discussion,"
from Vol. 3(2), 1992, 43–62.
    Chapter 1, from *Critical Essays, German Library, volume 98,* edited by Reinhold
Grimm and Bruce Armstrong. Translated by Stuart Hood. © 1982 by The
Continuum Publishing Company. Reprinted by permission. (Originally published in
the *New Left Review, 84,* 1974, 3–31.)
    Chapter 4, from *Socialism and the Limits of Liberalism,* edited by Peter Osborne.
© 1991 by Kate Soper. Reprinted by permission.
    Chapter 8, from the *New Left Review, 178,* 1989, 51–86. © 1989 by New Left
Books. Reprinted by permission.
    Chapter 11, excerpts from *Environmentalism and Political Theory,* by Robyn
Eckersley. © 1992 by State University of New York Press. Reprinted by permission.

INTRODUCTION TO THE DEMOCRACY AND ECOLOGY SERIES

This book series titled "Democracy and Ecology" is a contribution to the debates on the future of the global environment and "free market economy" and the prospects of radical green and democratic movements in the world today. While some call the post-Cold War period the "end of history," others sense that we may be living at its beginning. These scholars and activists believe that the seemingly all-powerful and reified world of global capital is creating more economic, social, political, and ecological problems than the world's ruling and political classes are able to resolve. There is a feeling that we are living through a general crisis, a turning point or divide that will create great dangers, and also opportunities for a nonexploitative, socially just, democratic ecological society. Many think that our species is learning how to regulate the relationship that we have with ourselves and the rest of nature in ways that defend ecological values and sensibilities, as well as right the exploitation and injustice that disfigure the present world order. Others are asking hard questions about what went wrong with the worlds that global capitalism and state socialism made, and about the kind of life that might be rebuilt from the wreckage of ecologically and socially bankrupt ways of working and living. The "Democracy and Ecology" series rehearses these and related questions, poses new ones, and tries to respond to them, if only tentatively and provisionally, because the stakes are so high, and since "time-honored slogans and time-worn formulae" have become part of the problem, not the solution.

JAMES O'CONNOR
Series Editor

ACKNOWLEDGMENTS

I should like to express my thanks to Peter Wissoker, of The Guilford Press, Barbara Laurence, of the Center for Political Ecology, and Jim O'Connor, the series editor, for helping in the various ways they did in the preparation of this book.

The introductory material and editorial work on this volume were made possible by the award of senior research fellowship (H52427505494) by the E.S.R.C.

CONTENTS

INTRODUCTION

At the end of the 1960s, there burst upon the world a storm of debate about impending environmental catastrophe. Paul Ehrlich's *The Population Bomb,* The *Ecologist*'s "A Blueprint for Survival," and the Club of Rome's *Limits to Growth*[1] were, in fact, quite different in their diagnoses and prognoses—but each in its own way signaled an irreversible break with the political past. The optimistic, modernizing expectation that technological advance and growing enlightenment would take us into a future of ever-rising living standards and greater human happiness was shattered. According to the formulation of the authors of *Limits,* continued exponential growth in population, industrial production, agricultural production, and pollution in a finite world must sooner or later overshoot its limits, with catastrophic consequences. The choices were stark: strong measures to restrict population growth and/or deep cuts in consumption and living standards.

Of course, critical voices were quick to point out that the "population bomb" was ticking primarily in the Third World, while the most profligate use of resources was taking place in the rich nations of the First World. The new politics of human ecology was easily open to the charge that it was little more than an updated and racialized version of Malthus's notorious "Law of Population." And the self-seeking character of calls to halt further industrial growth, while the great majority of the world's population remained in desperate need of the material benefits industrialism had brought in the West, was evident.

Nevertheless, the ecologists had a case that could not be *ignored;*

1

political ecology and the issues it addressed were a challenge to the whole
spectrum of Western social and political thought. Traditional debates
about liberty and democracy, rights and responsibility, equality and property, citizenship and authority had been overtaken by a new, insistent
question: the question of survival.

The challenge of the new ecological politics was addressed to Marxism
no less than to other traditions of political thought. Indeed, ecology posed
an arguably more profound challenge to the radical Left. Here was a new
critical perspective on the established social order, calling into question
premises Marxism had left untouched: the idea of historical progress to a
future "good life" founded on material abundance; the "progressive"
character of the advances of technology, even under capitalism; and so on.
Marxism's credentials as the most radical challenge yet to the status quo
were clearly threatened. Moreover, the newly recognized ecological crisis
demanded a theoretical response that Marxists initially were poorly prepared to give, despite the much proclaimed "materialism" of their intellectual tradition.

As it happened, the rise of ecological politics in the West coincided
with a remarkable, but quite distinct, renaissance in Marxist and neo-Marxist intellectual creativity. In some countries, such as France and Italy, which
had large and powerful communist parties, intellectual life was already
strongly colored by Marxist ideas. However, in the wake of the popular
movements of the 1960s in these countries, new spaces had been created
both for practical struggles and for the intellectual work of the Left. This
development was independent of, and often in conflict with, party leadership. In countries such as the United States and Great Britain, which lacked
strong communist parties (in the postwar period, at any rate), the renewal
of Marxist thought was primarily the work of intellectuals with relatively
little connection either to organized communism or to institutional politics
as such.

The result, in both settings, was a much more open-textured, self-criti-
cal, and receptive development of Marxist and neo-Marxist approaches,
one far less concerned with "orthodoxy" than in earlier phases of Marxism's history. This made it possible, at least, for creative dialogue to take
place between Marxists and non-Marxists, in the context of the new social
movements of liberation that emerged during the 1960s and 1970s,
especially those concerned with gender, sexuality, and racial oppression.
It was in this context, too, that Marxists began (or, as we shall see,
"recommenced") their response to the politics of ecology.

This book is a collection of writings that exhibit both the initial
diversity of these Marxist responses and their subsequent attempts to
reconstruct historical materialism itself, in ways informed by the newly
acknowledged centrality of the "metabolism" between historical societies

and the ecological conditions of their existence. Many, but by no means all, of these writings are reprinted from the pioneering international journal *Capitalism, Nature, Socialism (CNS)*. Part I consists of four selections (two of them drawn from *CNS*) by writers within, or strongly influenced by, the Marxist tradition taking stock of both the implications of ecological politics and the resources of their own tradition. Part II brings together four contributions to the ongoing quest for an ecologically informed historical materialism. Part III focuses on what many regard as the most systematically developed version of an ecological Marxism thus far: James O'Connor's thesis of a "second contradiction" in capitalism. His founding theoretical statement, published in the first issue of *CNS*, is reprinted here, with relatively minor revisions and four of the many commentaries published in subsequent issues of *CNS*. Finally, to continue in the spirit of critical engagement without which "ecological Marxism" could never have been born, the collection ends with a small selection of critical responses. Part IV includes two responses to the development of ecological Marxism from feminists (both previously published in *CNS*), together with a more extended response from a leading theorist of "ecocentric" politics. Each of these four parts of the book opens with a more detailed introduction to the ideas and arguments of the included essays.

NOTE

1. Paul Ehrlich, *The Population Bomb* (New York: Ballantine, 1970); Edward Goldsmith et al., "A Blueprint for Survival," *Ecologist* 2(1), January, 1972; and D. H. Meadows et al., *The Limits to Growth* (New York: Universe Books, 1972).

Marxism For
or Against Ecology?

INTRODUCTION
TO PART I

As the countercultural movements of the 1960s led to growing working-class militancy, especially in Europe, in the 1970s, Marxist thought and politics underwent a remarkable revival. Marxists, no less than the adherents of other political traditions, felt compelled to respond to the challenge of political ecology. Within the Marxist camp, there were three broad positions. The first, suspicious, skeptical, and sometimes outright hostile, was that the new politics of ecology was reactionary in content and elitist in terms of the interests it represented. Insensitive to the legitimate aspirations of the poor and dispossessed for a better standard of life, postulating a "universal interest" as a mask for the interests of the dominant classes, political ecology was nothing more nor less than a rejigged version of the old ruling ideology. Many Third World Marxists were—if only relatively briefly—tempted by this sort of response.

The second response to the challenge of political ecology was to dig back into the writings of Marx and Engels and to argue that, after all, they were the first political ecologists, and that the new prophets of ecology are telling us nothing we didn't know already. The third response was to acknowledge the significance, if not of the political ecologists themselves, then of the issues they raised. However, environmental crisis was to be seen not as the result of "industry" or "population," but as a consequence of the specifically capitalist form of organization of economic life. Seen in this way, political ecology confirmed, while extending and complementing, the

7

Marxist view of capitalism as its own grave digger. What the working class had so far failed to achieve might now be brought about by a rebellion of nature itself.

Of course, this threefold classification is something of a caricature. Most Marxists who took these issues seriously developed for more sophisticated positions, sometimes drawing on strands from all three of the above responses, and often displaying considerable ambivalence. Part I of this collection contains a tiny selection (four) of these more sophisticated responses to the ecological challenge, which attempt both to situate and to reassess the Marxian heritage in the face of the new ecological politics.

H. M. Enzensberger's "Critique" was one of the earliest and most influential Marxist responses, and has since become a classic. A large part of Enzensberger's text is given over to developing and endorsing the Left's critique of the new politics of ecology as ideology. This work of critique and clarification, he insists, is quite necessary. The extension of "ecology" as a subdiscipline of biology to the human case, and the subsequent extrapolations of global growth trends to demonstrate impending catastrophe, are full of methodological uncertainties and theoretical confusions. No one can ignore these prophecies because they have dramatic implications, but there is no way of knowing how near they are to the truth.

Later in his essay, however, Enzensberger seems to shift his position. Now the global ecological threat is acknowledged, and the most substantial objection to the ecological literature put forth is its lack of historical and sociopolitical content. Its global projections, and the metaphor "Spaceship Earth" function ideologically to conceal the massive disparities of material conditions of life, pollution, and resource use across the world, and the "structural dependency" of the poorer economies. The political solutions offered in this literature are correspondingly naive, unrealistic, and self-serving.

But the work of ideology critique does not stop at exposing the false or distorted character of a discourse: the further task is to explain it. Partly, for Enzensberger, this is a matter of displaying the social interests represented in the discourse of political ecology, but also it is necessary to show *why* such a high-profile debate about human ecology should be taking place *now*. As to the first task, Enzensberger offers a tripartite classification of the components of the ecology movement: a technocratic elite in the service of the ruling class; elements of the "new and old petit bourgeoisie" who campaign on specific, local environmental threats; and elements of the "reduced petit bourgeoisie"—remnants of the old "hippy" movement, practitioners of alternative life-styles, and so on.

Enzensberger is dismissive of this last group, but clearly takes the second group very seriously because they have the potential to become a genuine mass movement. Their turn to environmental protest and political action is a consequence of real changes in capitalist industrialization. Of course, environmental destruction itself is nothing new: it was an inseparable accompaniment of capitalist industrialization everywhere in the West. What is new is that conditions and threats formerly suffered by the working class are increasingly being experienced by the intermediate classes, who now find it either too difficult or too expensive to avoid them. Enzensberger here prefigures a central theme in later sociological writing about the "risk society."

However, the "real capitalists" *can* still escape environmental deterioration, so another explanation needs to be found for the ecological outpourings of their ideological representatives, the environmental technocrats. In part, this can be understood as a replay of the Malthusian controversy of early industrialization, though this time with a racist and imperialist emphasis on population growth rates in Third World countries. But here Enzensberger signals the need to go beyond the critique of ideology—this is, after all, simply to offer an "interpretation of an interpretation" that leaves the real conditions addressed by the original ideology untouched. In other words, Marxism itself needs to be developed to provide its own independent account of the processes alluded to, albeit ideologically, by the environmentalist discourse.

Enzensberger's attempt to do this himself can now be seen to be seriously flawed, but it is nonetheless full of suggestive and seminal insights. As we have seen, Enzensberger's view is that capitalist industrialization has always been ecologically destructive. He addresses the further thesis that as the *productive* powers developed under capitalist production relations have grown, so too have the *destructive* powers of this mode of production. This process has today reached the stage at which the material and human bases of capitalist production itself are threatened. But Enzensburger is not satisfied with this very general explanation for the newly emergent ecological concerns of capital itself. As well as constituting a *threat* to capital, environmental destruction also represents an opportunity for it. As the environmental costs of the production of commodities are offloaded onto the public and the state, new capitalist opportunities are created for profitable investment in environmental repair. Sections of capital acquire an interest in the recognition of environmental problems, giving rise to what Enzensberger calls an "eco–industrial complex."

Enzensberger's work makes a start on dissecting the political economy of environmental degradation and ecological politics. It insists that environmental destruction is endemic to capitalist industrialization, and offers a class analysis of the contradictory discourses and social movements that arise in response to this destruction. His account suggests that environmental problems can pose threats but also offer opportunities to different fractions and sectors of capitalist industry, so that we should expect conflicts of interest *within* the capitalist class over these issues. His analysis also clearly exposes the polyvalence of environmental discourses and social movements in terms of the contradictory interests of their diverse social bases. The importance of all these themes has been borne out by subsequent events, and they have been developed, as we shall see, by other ecological Marxist writers.

However, Enzensberger's attempt to provide a Marxist analysis of environmental degradation and ecological politics also brings him up against the limits of the inherited Marxist tradition itself. Here, clearly stated, is a recognition that the ecological challenge requires significant corrections and reworking of that heritage. These corrections concern three areas of Marxist thought and practice:

1. The role of "material progress." For Enzensberger this was linked to nineteenth-century technical optimism, which the founders of the materialist view of history shared with their "bourgeois" opponents. The fact that twentieth-century revolutions have characteristically taken place in "underdeveloped" countries shows that there is no essential connection between the "ripeness" of the developed forces of production and the transition to socialism.

2. The relation of forces and relations of production. Quoting another pioneer writer on the ecology–socialism connection, André Gorz,[1] Enzensberger questions the orthodox Marxist idea that socialism can simply take over, and thereafter use as the material base of the new society, the productive forces developed under capitalism. Capitalist technology embodies the imperatives of capitalist production relations and so cannot be regarded as a neutral or even as a liberatory achievement, requiring only those acts that would free it from capitalist "fetters." Enzensberger's own analysis of the symmetrical development of both productive and destructive powers with capitalist industrialization carries the same implication. The new society will require a new technical basis.

3. Survival versus abundance. To the extent that the ecological prophets of doom are right about the scale and irreversibility of environmental damage, it is clear that the relative wealth of the Western societies has been won at the expense of both Third World peoples and environments, and of future generations. The price of the long-delayed socialist revolution in the West is, therefore, that socialists can no longer hold out the promise of a future society of abundance. The future too belongs to the realm of necessity, and humankind is now faced with the problem of survival, not the hope of abundance.

This early ecological socialist equation of socialism and survival was a widespread response on the part of those socialists who were convinced that continued capitalist growth was leading to global ecological catastrophe. Rudolph Bahro,[2] an important early influence on the German Green Party, had already taken up this theme, arguing that environmental destruction in the West and in the centrally planned economies of Eastern Europe were both fueled by the global dynamic of capitalist accumulation and consumerism. Enzensberger, too, notes the significance of ecological damage in societies no longer characterized by private ownership of the means of production. For him, as for Bahro, this is not evidence against the thesis that socialism provides the solution to ecological destructiveness. A mere change in legal ownership, from private to state, that does not abolish the opposition between exchange value and use value, the fetishism of commodities, and the alienation of labor, cannot overcome the destructive dynamic of capitalist development.

Of course, this early perception of a link between socialism and ecological survival remained problematic in a number of ways. It implicitly accepted the thesis, borrowed from the much-maligned neo-Malthusian technocrats, that the projection of current growth trends spelled ecological disaster. All that was necessary in order to turn this into an argument for socialism was to link the growth dynamic with capitalism. But might a "steady-state" capitalism be possible? Even if the answer to this question is no, there is still the further question: Might not a regulatory regime be established that would render capitalist growth itself sustainable? And, of course, it is one thing to point out that capitalism is the problem, but quite another to demonstrate that socialism is the solution. *How* could socialism so order human relations to their environments to enable the meeting of need without environmental destruction? Clearly, much more needed to be said on the possible institutional forms that might be established in such a future society.

Finally, many questions still remained concerning the form of any collective agency that might be expected to bring about such changes. Enzensberger himself only speculates on the possibility—by no means the certainty—that a mass movement of the future might come to see the links between growing global destruction and the prevailing mode of production. These and other questions have preoccupied commentators on environmental politics—Marxist and non-Marxist alike—ever since, and the remaining sections of this book contain some significant contributions to the ensuing debates.

Clearly, Enzensberger's work transcends my initial three-fold classification of Marxist responses to the new ecological politics in that it embarks on the long struggle to reassess and revise the Marxist heritage. However, his work does combine elements of the first and third responses: critique of political ecology as ideology, and deployment of ecological arguments to complement the traditional Marxist case against capitalist economic and social organization. Enzensberger has little time for the more scholarly work of returning to the classics of Marxism in the hope of finding previously unrecognized sources of illumination on ecological questions.

Not surprisingly, given the textual–exegetical turn of much Western Marxism in the 1970s (especially under the influence of the Althusserian practice of "symptomatic" reading, and the more generalized "crisis of Marxism"), this latter response was more common than Enzensberger's direct engagement with the contemporary issues. The attempt to reinterpret the classic literature of Marxism in the light of current ecological concerns has continued since then, and Part I of this collection continues with three contrasting contributions by Jean-Guy Vaillancourt, Michael Perelman and Kate Soper.

Vaillancourt offers us a very balanced and informative review of some of the key texts written by Marx and Engels that address questions that today would be recognized as ecological. Engels's *Condition of the Working Class in England* is rightly praised as an important forerunner of human or social ecology, while Marx, especially in *Capital*, was fully aware of the simultaneously socially and environmentally destructive character of capitalism, especially in relation to capitalist agriculture. Vaillancourt brings together the much-discussed disputes over Malthus's law of population, Marx and Engels's shifting relationship to Darwinism, and Engels's later engagement with contemporary work in the natural sciences, to show a continuing concern with ecological issues on the part of the founders of Marxism. This review of their work shows that, widespread views to the contrary, there is no incompatibility between Marxism and ecology. On the other hand, it could not be claimed that Marx and Engels were the founders of modern human ecology. Their writings contain numerous ambiguities and they tended to oscillate between an anthropocentric commitment to the rational and scientific use of nature for human purposes, on the one hand, and, on the other, a more naturalistic materialism emphasizing human dependence on nature's laws and inherent dialectic.

Perelman's essay starts out from the widely held view that Marx failed to understand the economic importance of natural resources and so failed to anticipate the subsequent ecological problems of capitalism that are now of concern to economists. On Perelman's reading, Marx's *Capital does* have a sophisticated theory of natural resource scarcity. However, this is concealed by the use of misleading economic categories because of Marx's fear of giving ground to politically unacceptable Malthusianism. Perelman draws attention to a wide range of published and unpublished writings that demonstrate Marx's concern with the economic importance of resource scarcity, especially during and after the cotton shortage of the early 1860s caused by the U.S. Civil War. In general, Perelman argues, Marx deals with questions of "overpopulation," natural resource scarcity, rising food prices, and so on not in a Malthusian way as consequences of abstract general laws, but as historically specific effects of capitalist social organization in relation to its natural conditions.

More specifically, Marx's view was that capitalist agriculture was incapable of fully mastering organic processes in the way that capitalist industry had mastered inorganic ones. As the predominance of industrial over agricultural capitalism advances, so the demand for organic raw materials outstrips their supply. What Marx saw as an increasing difficulty in the supply of raw materials in line with capitalist industrial development accounted for the destruction of the fertility of the soil, and had consequences for the general rate of profit.

Marx's politically motivated reluctance to get dragged into further debate on Malthusian territory led him, Perelman argues, into adopting the concept of "constant capital" as a surrogate for raw materials in his discussion of the rising organic composition of capital, and the consequent tendency of the rate of profit to fall. The weakness of Marx's argument in this chapter of *Capital* can then be read as symptomatic of an unarticulated link, in Marx's mind, between (socially produced) resource scarcity and the rate of profit. Once rid of Marx's confused and confusing concealment of resource scarcity under the concept of the organic composition of capital, Perelman argues that Marx's general strategy for dealing with questions of resource scarcity, ecological degradation, and population growth has a great deal to offer contemporary ecological politics. In particular, the demonstration that each of these problems results from a particular relation of social organization to "nature" points to social and political change in response to ecological problems, and away from the illusory "technical fix."

Finally, to end Part I, Kate Soper returns to some classic texts of Marx—most especially the *Grundrisse*—to seek out the sources of green hostility to Marxism. She identifies three main grounds upon which many Greens object to Marxism: the "productivism" central to the Marxist account of history and vision of the future; the widespread identification of Marxism with the ecologically disastrous state-centralist regimes of Eastern Europe; and the Marxist commitment to the industrial working class as the main agent of change. Soper is clear that there is, indeed, much in the Marxist heritage that Greens are right to object to. Equally, however, she insists on the importance of not overlooking areas of consonance between red and green perspectives: the ecological dimension in Marx's early concept of "alienation," the shared red and green objection to a society in which all values are subordinated to money and profits, and the ways in which both the human exploitation and the environmental destruction involved in production are hidden from the view of the capitalist consumer.

Soper turns to Marx's philosophy, however, for a more positive evaluation of the potential contribution of Marxism to ecological politics. The dialectical ontology of labor, most evident in Marx's early works, offers an alternative to the polar opposition between "essentialist" green and Enlightenment views of the relation between humanity and nature. Though Marx's early writings are ambiguous, they do tend toward a view of humans as historically creating themselves through actions that transform their environment, but in the process also evoke new sensibilities and needs in the human subject. But this view of transformative potential is still set within outer limits imposed by nature and human nature itself.

Marx's later drafts for *Capital*, the *Grundrisse*, seem, however, to depart from this recognition of outer limits, to a more open-ended and Promethean vision of "mastery" of nature. This dimension in Marx's thinking, Soper concedes, is harder to reconcile with recent green thinking. Nevertheless, even the *Grundrisse* contains some important and still-relevant insights. In particular, Marx does not share the hostility to science and technology that is present in some green political circles, but he still fully acknowledges their destructive effects when set to work under modern capitalist social relations. Soper insists that sophisticated techniques will be needed in any ecological and socialist future to correct the depredations of the past. Clearly, the vision here is that of an actively, if benevolently, managed environment, as against a more contemplative respect for nature's autonomy. Soper's analysis coincides with Vaillancourt's reading of Marx and Engels as "Benedictine" rather than "Franciscan" in their approach to nature.

Finally, in Soper's account, Marx's distinction between necessary and surplus labor (which she associates with the distinction between necessary goods and luxuries) is of great value in opening up the possibility of a future in which work—something essentially bound up with resource use—might give way to freely creative and self-fulfilling forms of activity. As she points out, some of the activities Marx mentions, such as hunting and fishing, may not meet with green approval, but Marx's refusal to "blueprint" for humanity's future leaves room for his legacy to be developed in green directions.

Taken together, the essays collected in Part I reveal an equivocal relation between the Marxist Left and the new politics of ecology. Both share the recognition that the issues addressed by Greens are of fundamental—perhaps even overriding—importance, yet there is criticism of technocratic environmentalism as ideology, and of radical green "essentialism" about nature and human nature. The attempt to reread the Marxist classics with ecological questions in mind uncovers an array of lost or half-forgotten insights, and demonstrates the great potential value of the historical materialist tradition as a resource for political ecology. But at the same time as these insights are uncovered, it is recognized that the Marxist legacy *is* an ambiguous and uneven one. The great challenge remains: to critically reassess the Marxian heritage in the face of today's problems, and to further develop what remains defensible in it. In their different, even contrasting ways, that is the task embarked upon in the contributions to Part II of this book.

NOTES

1. See in particular André Gorz, *Ecology as Politics* (London: Pluto, 1983); and *Capitalism, Socialism, Ecology* (London: Verso, 1994).
2. See Rudolph Bahro, *Socialism and Survival* (London: Heretic, 1982); and *From Red to Green* (London: Verso, 1984).

Chapter 1

A CRITIQUE
OF POLITICAL ECOLOGY

HANS MAGNUS ENZENSBERGER

As a scientific discipline, ecology is about 125 years old. The concept emerged in 1868 when the German biologist Ernst Haeckel, in his *Natural History of Creation,* proposed giving this name to a subdiscipline of zoology—one that would investigate the totality of relationships between an animal species and its inorganic and organic environment. Compared with the present state of ecology, such a proposal suggests a comparatively modest program. Yet none of the restrictions contained in it proved to be tenable: neither the preference given to animal species over plant species, nor to macro- as opposed to microorganisms. With the discovery of whole ecosystems, the perspective that Haeckel had in mind became obsolete. Instead there emerged the concept of mutual dependence and of a balance between all the inhabitants of an ecosystem, and in the course of this development the range and complexity of the new discipline have grown rapidly. Ecology became as controversial as it is today only when its adherents decided to include a very particular species of animal in their research: man. While this step brought ecology unheard-of publicity, it also prompted a crisis about its validity and methodology, the end of which is not yet in sight.

Human ecology is, first of all, a hybrid discipline. In it categories and methods drawn from the natural and the social sciences have to be used together, despite the fact that this creates complications. Human ecology tends to stuck in more and more new disciplines and to subsume them under its own research aims. This tendency is justified by its practitioners not on scientific grounds but on the grounds of the urgency of ecology's aims. Under the pressure of public debate ecology's statements in recent years have become more and more markedly prognostic. This "futurological deformation" was totally alien to ecology so long as it considered itself merely as a particular area of biology. But today this science has now laid claim to a total validity—a claim that it cannot make good. The more far-reaching its conclusions, the less reliable they are. Since no one can vouch for the accuracy of the enormous volume of material from every conceivable science on which human ecology's hypotheses are constructed, it must—precisely to the degree that it wishes to make global statements—confine itself to offering working syntheses. One of the best known ecological handbooks, *Population, Resources, Environment,* by Paul and Anne Ehrlich, deploys evidence from the following branches of sciences either implicitly or explicitly: statistics, systems theory, cybernetics, games theory, and prediction theory; thermodynamics, biochemistry, biology, oceanography, mineralogy, meteorology, and genetics; physiology, medicine, epidemiology, and toxicology; agricultural science, urban studies, and demography; technologies of all kind; and theories of society, sociology, and economics (the latter admittedly in a most elementary form). It is hard to describe the methodological confusion that results from the attempt at a synthesis of this sort. If one starts from this theoretical position there can, obviously, be no question of producing a group of people who are competent to deal with it. From now on ecology is marginally relevant to everyone; and this, incidentally, is what makes the statements in this chapter possible.

THE CENTRAL HYPOTHESIS

What till recently was a marginal science has within a few years become the subject of bitter and widely discussed controversies. This cannot be explained merely by the snowballing effect of the mass media. It is connected with the central statement made by human ecology—a statement that refers to the future and is therefore at one and the same time prognostic and hypothetical. This hypothesis can be formulated as follows: The industrial societies of this earth are producing ecological contradictions, which must in the foreseeable future lead to their collapse. On the one hand, everyone is affected by this statement, since it relates to the continued existence of humanity; on the other hand, no one can form a clear and final judgment on it because, in the last analysis, it can only be verified or proved wrong in the future.

In contradistinction to earlier theories of catastrophe, this prognosis does not rest on linear, monocausal arguments. On the contrary, it introduces several synergetic factors. A very simplified list of the different strains of causality would look something like this:

1. Industrialization leads to an uncontrolled growth in world population. Simultaneously the material needs of that population increase. Thus, even given an enormous expansion in industrial production, the chances of satisfying human needs deteriorate per capita.

2. The industrial process has up to now been nourished from sources of energy that are not in the main self-renewing; among these are fossil fuels as well as supplies of fissile material like uranium. In a determinable space of time these supplies will be exhausted; their replacement with new sources of energy (such as atomic fusion) is theoretically conceivable, but not yet practically realizable.

3. The industrial process is also dependent on the employment of mineral raw materials—above all, metals—that are not self-renewing either; their exploitation is advancing so rapidly that the exhaustion of deposits can be foreseen.

4. The water requirements of the industrial process have reached a point where they can no longer be satisfied by the natural circulation of water. As a result, the reserves of water in the ground are being attacked; this must lead to disturbances in the present cycle of evaporation and precipitation and eventually to climatic changes. The only possible solution is the desalinization of seawater; but this process is so energy-intensive that it would accelerate the process described in Point 2 above.

5. A further limiting factor is the production of foodstuffs. Neither the area of land suitable for cultivation nor the yield per acre can be arbitrarily increased. Attempts to increase the productivity of farming lead, beyond a certain point, to new ecological imbalances, for example, to erosion, pollution through poisonous substances, and reductions in genetic variability. The production of food from the sea comes up against ecological limits of another kind.

6. A further factor—but only one factor among a number of others—is the notorious "pollution" of the earth. This category is misleading insofar as it presupposes a formerly "clean" world. This has naturally never existed, and is moreover ecologically neither conceivable nor desirable. What is actually meant are disequilibrium and dysfunctionings of all kinds in the metabolism between nature and human society occurring as the uninten-

tional side effects of the industrial process. The polycausal linking of these effects is of unimaginable complexity. Environmental poisoning caused by such harmful substances as pesticides, radioactive isotopes, detergents, pharmaceutical, food additives, artificial manures, lead and mercury, fluoride, and a vast quantity of other substances are only one facet of the problem. Irreversible waste is another facet of the same problem. Alterations in the atmosphere and in the resources of land and water traceable to metabolic causes such as production of smog, changes in climate, irreversible changes to rivers and lakes, and oceanographic changes must also be taken into account.

7. Scientific research into yet another factor does not appear to have got beyond the preliminary stages. There are no established critical quantifications for what is called "psychic pollution." Under this heading could be listed increasing exposure to excessive noise and other urban irritants, the psychological effects of overpopulation, and other stress factors that are difficult to isolate.

8. A final critical limit is presented by "thermal pollution." The laws of thermodynamics show that, even in principle, this limit cannot be crossed. Heat is emitted by all processes involving the conversion of energy. The consequences for the global supply of heat have not been made sufficiently clear.

A basic difficulty in the construction—or refutation—of ecological hypotheses is that the processes invoked do not take place serially but rather in close interdependence. That is also true of all attempts to find solutions to ecological crises. It often—if not always—emerges that measures to control one critical factor cause another to go out of control. One is dealing with a series of closed circuits, or rather of interference circuits, that are in many ways linked. Any discussion that attempted to deal with the alleged "causes" piecemeal and to disprove them simply would miss the core of the ecological debate and would fall below the level that the debate has in the meantime reached.[1]

Yet even if there exists a certain, but no by means complete, consensus that the present process of industrialization must lead ceteris paribus to a worldwide environmental breakdown, three important questions connected with the prognosis are still open to debate. The first concerns the timescale involved. Estimations of the point in time at which a galloping deterioration of the ecological situation may be expected differ by a magnitude of several centuries. They range from the end of the 1890s to the twenty-second century. In view of the innumerable variables involved in the calculations, such divergencies are not to be wondered at. For example, the critics of the MIT report *The Limits of Growth* have objected to the results given there on the grounds that the mathematical model on which it is based is much too simple and that the number of variables is too limited. A second controversial point is closely related to the first, namely, that the relative weight to be given to the individual factors that are blamed for the predicted catastrophe is not made clear. This is a point at issue, for example, in the debate between Barry Commoner and Paul Erlich. While the latter considers population growth to be the "critical factor," the former believes that the decisive factor is modern industrial technology. An exact analysis of the factors involved comes up against immense methodological difficulties. The scientific debate between the two schools therefore remains undecided.

Third, it is obviously not clear what qualifies as an "environmental catastrophe." In this connection one can distinguish a number of different perspectives dictated by expectation or fear. There are ecologists who concern themselves only with mounting dangers and the corresponding physiological, climatic, social, and political "disturbances" associated with them; and there are others, like the Swedish ecologist Gösta Ehrensvärd, who contemplate the end of social structures based on industrialization. Meanwhile, those who in the United States are called "doomsters" talk of the dying out of the human species or the disappearance from the planet of all primates, all mammals, or even all vertebrates. The tone in which the respective ecological hypotheses are presented ranges correspondingly from the mildest reformist warnings to deepest pessimism. What is decisive for the differences between them is the question of how far the process of ecological destruction and uncontrolled espolitation is to be regarded as irreversible. In the literature, the answer to this question depends either on an analysis of the factors involved, or on the temporal parameters. The uncertainty that is admitted to prevail on these two points means that there is no prospect of a firm answer. Authors like Ehrensvärd, who start from the premise that the end of industrial societies is at hand, and are already busy with preparations for a postindustrial society—one which, it should be added, contains a number of utopian traits—are still in the minority. Most ecologists do imply that they consider the damage done so far as reversible, if only by tacking on to their analyses proposals to avert the catastrophe of which they are the prophets. These proposals will need to be critically examined.

THE ECOLOGICAL MOVEMENT

Ecology's hypotheses about the future of industrialization have been disseminated, at least in the industrialized capitalist countries, through the mass media. The debate on the subject has itself to some extent acquired a mass character, particularly in the Anglo-Saxon and Scandinavian countries. It has led to the rise of a wide, although loosely organized, movement whose political potential is hard to estimate. At the same time the problem under discussion is peculiarly ill-defined. Even the statements of the ecologists themselves alternate between the construction of theories and broad statements of *Weltanschauung*, between precise research and totalizing theories linked to the philosophy of history. The thinking of the ecological groups often appears to be obscure and confused. The very fact that it is disseminated by the mass media means that the debate generally loses a great deal of its stringency and content. Subordinate questions such as that of recycling refuse or "pollution" are treated in isolation; hypotheses are presented as unequivocal truths; spectacular cases of poisoning are sensationally exploited; isolated research results are given absolute validity; and so on. Their processing through the sewage system of industrialized publicity has therefore, to some extent, led to further pollution of a cluster of problems that from the start could not be presented in a "pure" way. This lack of clarity is propagated in the groups that are at present actively occupied with the subject of ecology, or rather with its *disjecta membra*, with what is left of it. The most powerful of these groups is that of the technocrats who, at all levels of the state apparatus and also of industry, are busy finding the speediest solutions to particular problems—"quick technological fixes"—and then implementing them. This they do whenever there is a considerable potential for economic or political conflict—and only then. These people consider themselves to be entirely pragmatic—that is to say, they are servants of the ruling class at present in power—and cannot be assumed to have a proper awareness of the problem. They can be included in the ecological movement only insofar as they belong—as will be demonstrated—to its manipulators and in so far as they benefit from it. The political motives and interests in these cases are either obvious—as with the Club of Rome, a consortium of top managers and bureaucrats—or can easily and unequivocally be established.

What is less unequivocal is the political character of a second form of ecological awareness and the practice that corresponds to it. I am speaking of the many smaller groups of "concerned and responsible citizens," as they say in the United States. The expression points, as does its German parallel, "citizens' initiative," to the class background of those involved in it. They are overwhelmingly members of the middle class and of the new petite bourgeoisie. Generally, their activities have modest goals. They are concerned with preserving open spaces or trees. They encourage classes of school children to clean up litter on beaches or recreation grounds. They organize a boycott of nonbiodegradable packaging. The harmless impression made by projects of this kind can easily blind us to the reserves of militancy they conceal. There only needs to be a tiny alteration in the definition of goals and these groups spontaneously begin to increase in size and power. They are then able to prevent large-scale industrial projects like the siting of an oil refinery, to force utility companies to relocate high-tension cables underground, or to pressure the government to cancel the construction of a planned highway. But even achievements of this magnitude only represent the limits of their effectiveness for a time. If the hypotheses of the ecologists should come even partially true, today's small and disorganized ecological action groups will become a force of the first order in domestic politics and one that can no longer be ignored. On the one hand, they express the powerful and legitimate needs of those who engage in these activities; on the other hand, they set their sights on immediate targets, which are not understood politically, and incline to a kind of indulgence in social illusion. This makes them ideal fodder for demagogues and interested third parties. But the limited nature of their initiatives should not conceal the fact that there lies within them the seed of a possible mass movement.

Finally, there is that part of the ecological movement that considers itself to be its hard core but which, in fact, plays a rather marginal role. These are the "ecofreaks." Those groups, which have mostly split off from the American protest movement, are engaged in a kind of systematic flight from the cities and from civilization. They live in rural communes, grow their own food, and seek a "natural way of life," which may be regarded as the simulation of pre- or postindustrial conditions. They look for salvation in detailed, precisely stipulated dietary habits—eating "earth food"—and agricultural methods. Their class background corresponds to that of the hippies of the 1960s—of reduced middle-class origin, enriched by elements from peripheral groups. Ideologically they incline toward obscurantism and sectarianism.

On the whole one can say that in the ecological movement—or perhaps one should say "movements"—the scientific aspects, which derive predominantly from biology, have merged in an extremely confused alliance with a whole series of political motivations and interests, which are partly manifest, partly concealed. At a deeper level one can identify a great number of sociopsychological needs, which are usually aroused without those concerned being able to see through them. These include hopes of conversion and redemption, delight in the collapse of things, feelings of guilt and resignation, escapism and hostility to civilization.

In these circumstances, it is not surprising that the European Left holds itself aloof from the ecological movement. It is true that it has incorporated certain topics from the environmental debate in the repertory of its anticapitalist agitation, but it maintains a skeptical attitude to the basic hypothesis underlying ecology and avoids entering into alliances with groups that are entirely oriented toward ecological questions. The Left has instead seen its task to be to face the problem in terms of an ideological critique. It therefore functions chiefly as an instrument of clarification, as a tribunal that attempts to dispel the innumerable mystifications that dominate ecological thinking and have encouraged it. The most important elements in this process of clarification, which is absolutely necessary, are listed and discussed below.

THE CLASS CHARACTER
OF THE CURRENT ECOLOGICAL DEBATE

The social neutrality to which the ecological debate lays claim, leaving recourse as it does so to strategies derived from the evidence of the natural sciences, is a fiction. A simple piece of historical reflection shows just how far this class neutrality goes. Industrialization made whole towns and areas of the countryside uninhabitable as long as 150 years ago. The environmental conditions at places of work, that is to say, in the English factories and pits, were—as innumerable documents demonstrate—dangerous to life. There was an infernal noise. The air people breathed was polluted with explosive and poisonous gases, as well as with carcinogenic matter and particles that were highly contaminated with bacteria. The smell was unimaginable. In the labor process contagious poisons of all kinds were used. The workers' diet was bad. Food was adulterated. Safety measures were non-existent or were ignored. The overcrowding in the working-class quarters was notorious. The situation regarding drinking water and drainage was terrifying, as the following quotation indicates:

> When cholera prevailed in that district [Tranent, in Scotland] some of the patients suffered very much indeed from want of water, and so great was the privation, that on that calamitous occasion people went into the ploughed fields and gathered rain water which collected in depressions in the ground, and actually in the prints made by horses' feet. Tranent was formerly well-supplied with water of excellent quality by a spring above the village, which flows through a sand-bed. The water flows into Tranent at its head and is received into about ten wells, distributed throughout the village. The people supply themselves at these wells when they contain water. When the supply is small, the water pours in a very small stream only. . . . I have seen women fighting for water. The wells are sometimes frequented throughout the whole night. It was generally believed by the population that this stoppage of the water was owing to its stream being diverted into a coal-pit which was sunk in the sand-bed above Tranent.[2]

These conditions, which are substantiated by innumerable other sources from the nineteenth century, would undoubtedly have presented a "neutral observer" with food for ecological reflection. But there were no such observers. It occurred to no one to draw pessimistic conclusions about the future of industrialization from these facts. The ecological movement has only come into being since the districts that the bourgeoisie inhabit have been exposed to those environmental burdens that industrialization brings with it. What fills their prophets with terror is not so much ecological decline, which has been present since time immemorial, as its universalization. To isolate oneself from this process becomes increasingly difficult. It deploys a dialectic that in the last resort turns against its own beneficiaries. Pleasure trips and expensive packaging, for example, are by no means phenomena that have emerged only in the last decades; they are part of the traditional consumption of the ruling classes. They have become problematic, however, in the shape of tourism and the litter of consumerism, that is, only since the laboring masses have begun to share them. Quantitative increase tips over into a new quality, that of destruction. What was previously privilege now appears as nightmare and capitalist industry proceeds to take tardy, if still comparatively mild, revenge on those who up to now had only derived benefit from it. The real capitalist class, which is decreasing in numbers, can admittedly still avoid these consequences. It can buy its own private beaches and employ lackeys of all kinds. But for both the old and the new petite bourgeoisie such expenditure is unthinkable. The cost of a private "environment" that makes it possible to escape to some extent from the consequences of industrialization is already astronomical and will rise more sharply in future.

It is easy to understand why the working classes care little about general environmental problems and are only prepared to take part in campaigns when they have the opportunity to directly improve their own working and living conditions. Insofar as it can be considered a source of ideology, ecology is a matter that concerns the middle class. If avowed representatives of monopoly capitalism have recently become its spokesmen—as in the Club of Rome—that is because of reasons that have little to do with the living conditions of the ruling class. These reasons require analysis.

THE INTERESTS OF THE ECO-INDUSTRIAL COMPLEX

That the capitalist mode of production has catastrophic consequences is a commonplace of Marxism, which also not infrequently crops up in the arguments of the ecological movement. Certainly the fight for a "clean" environment always contains anticapitalist elements. Nevertheless, the rise of fascism in Germany and Italy demonstrated how easily such elements can be turned round and become tools in the service of the interests of capital.[3] It is therefore not surprising that ecological protest, at least in Western Europe, almost always ends up with an appeal to the state. Under present political conditions this means that it appeals to reformism and to technocratic rationality. This appeal is then answered by government programs that promise "improvement in the quality of life," without of course indicating whose life is going to be made more beautiful, in what way, and at whose expense. The state only "goes into action when the earning powers of the entrepreneur are threatened. Today the environmental crisis presents a massive threat to these interests. On the one hand it threatens the material basis of production—air, earth and water—while on the other hand it threatens man, the productive factor, whose usefulness is being reduced by frequent physical and mental illnesses."[4] To these have to be added the danger of uncontrollable riots over ecological questions as the environment progressively deteriorates.

On the question of state intervention and "environmental protection from above," the Left's ideological critique displays a remarkable lack of historical reflection. Here, too, it is certainly not a question of new phenomena. The negative effects of environmental damage on the earning power of industry, the struggle over the off-loading of liability, over laws relating to the environment, and over the range of state control can be traced back without much difficulty to the early period of English industrialization. A remarkable lack or variation in the attitude of the interests involved emerges from such a study. The previously quoted report on the water supply and the drainage problems in a Scottish mining village is taken from an official report of the year 1842—one that incidentally was also quoted by Engels in his book on *The Condition of the English Working Class.* The chairman of the commission of inquiry was a certain Sir Edwin Chadwick, a typical predecessor of the modern ecological technocrats. Chadwick was a follower of the utilitarian political philosopher and lawyer Jeremy Bentham, of whom Marx said: "If I had the courage of my friend H. Heine, I would call Mr. Jeremiah a genius at bourgeois stupidity."[5] James Ridgeway, one of the few American ecologists capable of intervening in the present environmental discussion with political arguments, has dealt thoroughly with Chadwick's role.[6] Then as now the rhetoric of the ecological reformers served to cloak quite concrete connections between a variety of interests. The technological means with which this "reform from above" operates have also altered less than one might think.[7]

But a historical perspective fails in its object if it is used to reduce modern problems to the level of past ones. Ridgeway does not always avoid this danger: he tends to restrict himself to traditional ecological questions like water pollution and the supply of energy. Without meaning to do so he thereby reduces the extent of the threatened catastrophe. It is true that there were environmental crises before our time and that the mechanisms of reformist managements set up to deal with such crises have a long history. What has to be kept in mind, however, is that ecological risks have not only increased quantitatively, but they have taken on a new quality.

In line with the changes that have taken place in the economic basis, this also holds true for environmental pollution and state intervention. In its present form monopoly capitalism is inclined, as is well known, to solve its demand problems by extravagant expenditure at the cost of the public treasury. The most obvious examples of this are unproductive investments in armaments and in space exploration. Industrial protection of the environment emerges as a new growth area the costs of which can either be off-loaded on to prices, or are directly made a social charge through the budget in the form of subsidies, tax concessions, and direct measures by the public authorities, while the profits accrue to the monopolies. "According to the calculations of the American Council of Environmental Quality at least a mission dollars is pocketed in the course of the elimination of three million dollars' worth of damage to the environment."[8]

Thus the recognition of the problems attendant on industrial growth serves to promote a new growth industry. The rapidly expanding eco-industrial complex makes profits in two ways: in the straightforward market, where consumer goods for private consumption are produced with increasing pollution, and in another where that same pollution has to be contained by control techniques financed by the public. This process at the same time increases the concentration of capital in the hands of a few international concerns, since the smaller industrial plants are not in a position to provide their own finance for the development of systems designed to protect the environment.

For these reasons the monopolies attempt to acquire influence over the ecological movement. The MIT study commissioned by the Club of Rome is by no means the only initiative of this kind. The monopolies are also represented in all state and private commissions on the protection of the environment. Their influence on legislation is decisive, and there are numerous indications that even apparently spontaneous ecological campaigns have been promoted by large firms and government departments. There emerges a policy of "alliances from above" whose demagogic motives are obvious.[9]

By no means all ecological movements based on private initiative put themselves at the service of the interests of capital with such servility. That is demonstrated by the fact that their emergence has often led to confrontations with the police. The danger of being used is, however, always present. It must also be remembered that the interests of capital contain their own contradictions. Ecological controversies often mirror the clash of interests of different groups of entrepreneurs without their initiators always being clear as to the stakes involved in the campaigns. A long process of clarification will be necessary before the ecological movement has reached that minimum degree of political consciousness that it would require to finally understand who its enemy is and whose interests it has to defend.[10]

DEMOGRAPHY AND IMPERIALISM

Warnings about the consequences of uncontrolled population growth—the so-called population explosion—also contain ideological motives. Behind the demands to contain it are concealed political interests that do not reveal themselves openly. The neo-Malthusian arguments that authors like Ehrlich and Taylor have been at pains to popularize found expression at a particular moment in time and in a quite particular political context. They originate almost exclusively from North American sources and can be dated to the late 1950s and early 1960s, a time when the liberation movements in the Third World began to become a central problem for the leading imperialist power. (On the other hand, the rate of increase in population had begun to rise much earlier, in the 1930s and 1940s.)

That this is no mere coincidence was first recognized and expressed by the Cubans:

> At that time [1962] the Population Council in New York, supported by the Population Reference Bureau Inc. in Washington, launched an extensive publicity campaign for neo-Malthusianism with massive financial help from the Ford and Rockefeller Foundations, which contributed millions of dollars. The campaign pursued a double goal, which may even be attained: the ruling classes of Latin America were to be persuaded by means of skillful propaganda based on the findings of the FAO and work done by numerous, even progressive scientists, that a demographic increase of 2.5 per cent in Latin America would lead to a catastrophe of incalculable dimensions. The following excerpts from the report of the Rockefeller Foundation for 1965 are typical of this literature *made in the USA*: "The pessimistic prediction that humanity is soon likely to be stifled by its own growth increasingly confronts all attempts to bring about an improvement in living standards. . . . It is clear that mankind will double in numbers in the lifetime of two generations unless the present growth tendency is brought under control. The results will be catastrophic for innumerable millions of individuals." The Population Reference Bureau expresses itself even more unequivocally: "The future of the world will be decided in the Latin American continent, in Asia and Africa, because in these developing territories the highest demographic rates of growth have been

registered. Either the birth rates must be lowered or the death rate must rise again if the growth is to be brought under control. . . . The biologists, sociologists and economists of the Bureau have forecast the moment when Malthus' theory will return like a ghost and haunt the nations of the earth" [P.R.B. press statement of October 1966].

The Cuban report also quotes Lyndon B. Johnson's remark to the effect that "five dollars put into birth control is more useful in Latin America than a hundred dollars invested in economic growth."[11] It then adds: "A comment on this cynical statement seems to us to be superfluous."

Indeed, not much intelligence is needed to discover behind the benevolent pose of the Americans both strong political motivations and the irrational fears that are responsible for the massive attempt by official and private groups in the United States to export birth control to the countries of the Third World. The imperialist nations see the time coming when their populations will be only a small minority when compared to the rest of the world population, and their governments fear that population pressures will become a source of political and military power. Admittedly, fears of another kind can be detected underneath the rational calculations: symptoms of a certain panic, the precursors of which are easily recognizable in history. One has only to think of the hysterical slogans of the heyday of imperialism (e.g., "The Yellow Peril") and of the period of German fascism (e.g., "the Red Hordes"). The "politics" of population have never been free of irrational and racist traits; they always contain demagogic elements and are always prone to arouse atavistic feelings. This is admittedly true not only for the imperialist side. Even the Cuban source does not stop at the extremely enlightening comment that has been quoted but continues as follows:

> Fidel Castro has spoken on the question many times. We recall his words now: "In certain countries they are saying that only birth control provides a solution to the problem. Only capitalists, the exploiters, can speak like that; for no one who is conscious of what man can achieve with the help of technology and science will wish to set a limit to the number of human beings who can live on the earth. . . . That is the deep conviction of all revolutionaries. What characterized Malthus in his time and the neo-Malthusians in our time is their pessimism, their lack of trust in the future destiny of man. That alone is the reason why revolutionaries can never be Malthusians. *We shall never be too numerous* however many of us there are, if only we all together place our efforts and our intelligence at the service of mankind, a mankind that will be freed from the exploitation of man by man."[12]

In such phrases not only does the well-known tendency of the Cuban revolution to voluntarism find expression together with a rhetoric of affirmation; but there is also the tendency to answer the irrational fears of the imperialist oppressor with equally irrational hopes. A materialist analysis of concrete needs, possibilities, and limits cannot be replaced by figures of speech. The Chinese leadership recognized that reality long ago and has therefore repeatedly modified its earlier population policy, which was very similar to the Cuban one in its premises. As far as the neo-Malthusians in the United States are concerned, a violent conflict has been raging for several years over their theses and their motivation.

THE PROBLEM OF GLOBAL PROJECTION

A central ideological theme of the ecological debate as it is at present conducted—indeed, it is at its very heart—is the metaphor of "Spaceship Earth." This concept belongs above all to the repertory of the American ecological movement. Scientific debates tend to sound more sober, but their content comes down to the same thing: they consider the planet as a closed and global ecosystem.

The degree of "false consciousness" contained in these concepts is obvious. It links up with platitudes that are considered "idealistic" but to which even that word is misapplied: "The good of the community takes precedence over the good of the individual," "We are all in the same boat," and so on. The ideological purpose of such hasty global projections is clear. The aim is to deny once and for all that little difference between first class and steerage, between the bridge and the engine room. One of the oldest ways of giving legitimacy to class domination and exploitation is resurrected in the new garb of ecology. Forrester and Meadows, the authors of the MIT report, for instance, by planning their lines of development from the start on a world scale, and always referring to the "Spaceship Earth"—and who would not be taken in by such global brotherliness?—avoid the need to analyze the distribution of costs and profits, to define their structural limitations, and with them the wide variation between the chances of bringing human misery to an end. For while some can afford to plan for growth and still draw profits from the elimination and prevention of the damage they do, others certainly cannot. Thus, under accelerated state capitalism, the industrial countries of the northern territories of the world can maintain capital accumulation by diverting it to antipollution measures, to the recycling of basic raw materials, to processes involving intensive instead of extensive growth. This path is denied to the developing countries, which are compelled to exploit to the utmost their sources of raw materials and, because of their structural dependence, are urged to continue intensive exploitation of their own resources. (It is worth quoting in this connection the remark of a Brazilian minister of economics to the effect that his country could not have enough pollution of the environment if that was the cost of giving its population sufficient work and bread.[13])

The contradictions that the ecological ideologies attempt to suppress in their global rhetoric emerge all the more sharply the more one takes their prognoses and demands at their face value. What would be the concrete effect, for instance, of a limitation of the consumption of energy over the whole of "Spaceship Earth" such as is demanded in almost all ecological programs?

Stabilization of the use of energy—certainly, but at what level? If the average per capita consumption of a United States citizen is to serve as a measure, then a future world society, stabilized at this level, would make an annual demand on the available reserves of energy of roughly 350×10^{12} kilowatt hours. The world production of energy would then be almost seven times as great as it is at present and the thermal, atmospheric, and radioactive pollution would increase to such a degree that the consequences would be unforeseeable; at the same time the available reserves of fossil fuel would disappear. If one chooses the present world average instead of the energy standard of the United States today as a measure of a future "stable" control of energy, then the exploitation of the available source of energy and the thermal, chemical, and radioactive effects in the environment would settle at a level only slightly higher than at present and one that would be tolerable in the long run. The real question would then be, however, how the available energy should be distributed globally. In arithmetical terms the solution would look something like this. The developing countries would have to have three times as much energy at their disposal as they do today; the socialist countries could by and large maintain their present level of consumption; but the highly industrialized countries of Europe and the United States would have to reduce their consumption enormously and enter upon a period of *contraction*.[14]

It must be clear that redistributions of such magnitude could be put through only by force: this is bound to hold good not only in international but also in national terms. Admittedly the captains of industry, gathered together in the Club of Rome, appear to have another view of conditions on board the spaceship in which we are supposed to be sitting. They are clearly not plagued by doubts as to their own competence and qualities of leadership. On the contrary, they assert that "very few people are thinking about the future from a global point of view."[15] This minority leaves no doubt that they are determined to adjust their view of the world to suit their own interests. The scarcer the resources, the more one adopts this view in distributing them; but the more one adopts this view of the world, the fewer people can be considered for this high office.

An ecologist who finds himself confronted by objections of this kind will generally attempt to counter them by changing the terms of the argument. He will explain that his immediate task is to deal with a condition that exists in fact; this is a task that takes precedence over future distribution problems that it is not his task to solve. On a factual level, however, it is impossible not to treat the problem on a global scale; indeed, it is inevitable. The pollution of the oceans or of the atmosphere, the spread of radioactive isotopes, the consequences of man-made changes in climate—all these are *actually,* and not merely in an ideological sense, worldwide and global phenomena and can be understood only as such.

While that is true, it does not help much. So long as ecology considered itself to be a branch of biology it was always conscious of the dialectical connection between the whole and the part; far from wishing "merely" to investigate life on earth, it saw itself as a science of interdependence and attempted to investigate the relations between individual species, the ecological subsystem in which they live and the larger systems. With the expansion of its research aims, its claims to hegemony, and the consequent methodological syncretism, human ecology has forfeited that ability to differentiate that characterized its founders. Its tendency to make hasty global projections is in the last analysis a surrender in the face of the size and complexity of the problem that it has thrown up. The reason for this failure is not difficult to determine. An ecologist researching the conditions of life in a lake has solid methodological ground to stand on; ecological arguments begin to become shaky only when the ecologist involves his own species in them. Escape into global projection is the simplest way out. For in the case of man, the mediation between the whole and the part, between subsystem and global system, cannot be explained by the tools of biology. This mediation is *social,* and its explication requires an elaborated social theory and at the very least some basic assumptions about the historical process. Neither the one nor the other is available to present-day ecologists. That is why their hypotheses, in spite of their factual core, so easily fall victim to ideology.

ENVIRONMENTAL APOCALYPSE AS AN IDEOLOGICAL PAWN

The concept of a critique of ideology is not clearly defined, nor is the object it studies. It is not only that "false consciousness" proliferates in extraordinary and exotic luxuriance given the present conditions under which opinions are manufactured, but it is also as consistent as a jellyfish and capable of protean feats of adaptability. So far I have examined the most widely diffused components of environmental ideology chiefly with regard to the interests they at once conceal and promote. This would have to be distinguished from an evaluation in terms of an ideological critique that sees the ecological debate as a symptom that yields conclusions about the state of the society that produces it. So that nothing may be omitted, interpretations of this kind will now be briefly surveyed, although it is doubtful whether this survey will bring to light any new perspectives.

From this point of view, the preoccupation with ecological crisis appears as a phenomenon belonging entirely to the superstructure—namely, an expression of the decadence of bourgeois society. The bourgeoisie can conceive of its own imminent collapse only as the end of the world. Insofar as it sees any salvation at all, it sees it only in the past. Anything of that past that still exists must be preserved, must be conserved. In earlier phases of bourgeois society this longing for earlier cultural conditions concentrated on "values" that either did obtain previously or were believed to have done so. With the progressive liquidation of this "inheritance," for example, religion, the search for the roots of things, which is now thought to reside in what is left of "nature," has become radicalized. In its period of decadence the bourgeoisie therefore proclaims itself to be the protector of something that it itself destroyed. It flees from the world that, so long as it was a revolutionary class, it created in its own image, and wishes to conserve something that no longer exists. Like the apprentice sorcerer it would like to get rid of the industrialization to which it owes its own power. But since the journey into the past is not possible, it is projected into the future: a return to barbarism, which is depicted as a preindustrial idyll. The imminent catastrophe is conjured up with a mixture of trembling and pleasure and awaited with both terror and longing. Just as, in German society between the wars, Klages and Spengler sounded the apocalyptic note, so in the Anglo-Saxon lands today the ecological Cassandras find a role as preachers calling a class that no longer believes in its own future to repentance. Only the scale of the prophecies has changed. While Klages and Spengler contemplated the decline of Europe, today the whole planet must pay for our hubris. Whereas in those days a barbarian civilization was to win terrible victories over a precious culture, today civilization is both victim and executioner. What will remain, according to the prophecies, is not an inner desert but a physical one. And so on. However illuminating such excesses may occasionally sound, they cannot advance beyond a point of view that is little more than that of the history of ideas. Besides, they do not carry much conviction in view of the fact that the dominant monopolies of the capitalist world show no signs of becoming aware of their presumed decadence. Just as German industry in the 1920s did not allow itself to be diverted from its expansion, so IBM and General Motors show little inclination to take the MIT report seriously. Theories of decline are a poor substitute for materialist analyses. If one explores their historical roots it usually emerges—as in the case of Lukács, that they are nourished by that very idealism that they claim to criticize.

THE CRITIQUE OF IDEOLOGY AS AN IDEOLOGY

The attempt to summarize the Left's arguments has shown that its main intervention in the environmental controversy has been through the critique of ideology. This kind of approach is not completely pointless, and there is no position other than Marxism from which such a critical examination of the material would be possible. But an ideological critique is only useful when it remains conscious of its own limitations: it is in no position to handle the object of its own research. As such it remains merely the interpretation of an interpretation of real conditions, and is therefore unable to reach the heart of the problem. Its characteristic gesture of "unmasking" can turn into a smug ritual if attention remains fixed on the mask instead of on what is revealed beneath it. The fact that we name the interests that lie behind current demographic theories will not conjure the needs of a rapidly growing population out of existence. An examination of the advertising campaigns of the enterprises involved does not increase the energy reserves of the earth by a single ton. And the amount of foreign matter in the air is not in any way reduced if we draw attention to the earlier history of pollution in the working-class quarters of Victorian England. A critique of ideology that is tempted to go beyond its effective limits itself becomes an ideology.

The Left in West Germany has so far been scarcely conscious of this danger, or at least has not thought about it adequately, although it is by no means new in historical terms. Even Marxist thinking is not immune to ideological deformations, and Marxist theory, too, can become a false consciousness if, instead of being used for the methodical investigation of reality through theory and practice, it is misused as a defense against that very reality. Marxism as a defensive mechanism, as a talisman against the demands of reality, as a collection of exorcisms—these are tendencies that we all have reason to take note of and to combat. The issue of ecology offers but one example. Those who wish to deprive Marxism of its critical, subversive power and turn it into an affirmative doctrine generally do so by using a series of stereotyped statements which, in their abstraction, are as irrefutable as they are devoid of results. One example is the claim announced in the pages of every other picture magazine, irrespective of whether it is discussing syphilis, an earthquake, or a plague of locusts, that "capitalism is to blame!"

It is splendid that anticapitalist sentiments are so widespread today that even glossy magazines cannot avoid expressing them. But it is quite another question how far an analysis deserves to be called Marxist when it a priori attributes every conceivable problem to capitalism, and what the political effect of this is. Its commonplace nature renders it harmless. Capitalism, so frequently denounced, becomes a kind of social ether, omnipresent and intangible, a quasi-natural cause of ruin and destruction, the exorcising of which can have a positively neutralizing effect. Since the concrete problem in hand—whether psychosis, lack of nursery schools, dying rivers, or air crashes—can, without precise analysis of the exact causes, be referred to the total situation, the impression is given that any specific intervention here and now is pointless. In the same way, reference to the need for revolution becomes an empty formula, the ideological mask of passivity.

The same holds true for the thesis that ecological catastrophe is unavoidable within the capitalist system. The prerequisite for all solutions to the environmental crisis is the introduction of socialism. No particular skill is involved in deducing this answer from the premises of Marxist theory. The question, however, is whether it adds up to more than an abstract statement that has nothing to do with political praxis and which allows whoever utters it to neglect the examination of his or her concrete situation.

The ideological packaging of such statements is dispelled at once, however, if one asks what exactly they mean. The mere question of what is meant by "capitalism" brings to light the most crass contradictions. The comfortable structure of the commonplace falls apart. What is left is a heap of unresolved problems. If one understands by "capitalism" a system characterized by private ownership of the means of production, then it follows that the ecological problem, like all the other evils of which "capitalism" is guilty, will be solved by nationalization of the means of production. It also follows that in the Soviet Union there can be no environmental problems. Anyone who asserts the contrary must be prepared to be insulted if he produces a bundle of quotations from *Pravda* and *Izvestia* about the polluted air of the Don Basin or the filthy Volga as evidence. Such a comparison of systems is forbidden—at least by Marxists like Gerhard Kade, who argues:

> For all those who are embarrassed by the question of the relationship between bourgeois capitalist methods of production and the destruction of the environment, a well-proven argument can be produced from that box of tricks where diversionary social and political tactics are kept. Scientists talk of comparing the two systems: standard commonplace minds immediately think of the filthy Volga, the polluted air of the Don Basin, or of that around Leuna. A whole tradition lies behind this tactic. There is no social or political issue, from party conferences to reports

on the state of the nation, where the diversionary effectiveness of such comparisons between systems has not already proved its worth. Whatever emerges from the increasing number of inquiries into environmental pollution in the socialist countries dressed up scientifically and becomes a useful weapon in a situation where demands for replacement of the system begin to threaten those who have an interest in upholding present conditions. "Go to East Germany if you don't like it here" or "Throw Dutschke over the Wall" are the socially aggressive forms adopted by that diversionary maneuver.[16]

Critique of ideology as ideology: the position that lays the blame on "capitalism" is defended here at the cost of its credibility. Moreover the fact that in the socialist countries the destruction of the environment has also reached perilous proportions is not even disputed, merely ignored. Anyone who is not prepared to go along with this type of scientific thinking is guilty of drawing analogies between the systems and is denounced as an anticommunist, a sort of ecological Springer. The danger that such a denatured form of Marxism will establish a hold on the masses is admittedly slight. The relationship of the German working class to its own reality is not so remote as to exclude the possibility of a comparative examination. In the face of such narrowness, one must

bear in mind that capitalism as a historical form and as a system of production cannot be identified with the existence of a class of owners. It is an all-embracing social mode of production arising from a particular type of accumulation and reproduction that has produced a network of relationships between human beings more complicated than any in the history of man. This system of production cannot simply be done away with by dispossessing private capitalists, even when this expropriation makes it possible in practice to render that part of surplus value available for other purposes that is not used for accumulation. The socialist revolution cannot be understood merely as a transfer of ownership leading to a more just distribution of wealth while other relationships remain alienated and reified. On the contrary, it must lead to totally revolutionized relationships between men and between men and things— that is to say, it must revolutionize the whole social production of their lives. It will either aim to transcend the proletariat's situation, of alienation, of the division between work and its profit, and the end of commodity fetishism, or it will not be the socialist revolution.[17]

Only such a view of capitalism as a mode of production and not as a mere property relationship, allows the ecological problem to be dealt with in Marxist terms. In this connection the categories of use value and exchange value are of decisive importance. The disturbance of the material interchange between man and nature is then revealed as the strict consequence of capitalist commodity production.[18] This is a conclusion that makes the ideological ban on thought unnecessary and explains why ecological problems survive in the socialist countries, too. After all, the contradiction between use value and exchange value is not superseded any more than wage labor and commodity production. "Socialist society has remained a transitional society in a very precise meaning of the word—a social form in which the capitalist mode of production, compounded with new elements, continues to exist and exercises a decisive pressure on the political sphere, on relations between human beings, and on the relationship between rulers and ruled."[19] No less decisive is the pressure that the persistence of the capitalist mode of production exercises on the relationship between man and nature—a pressure that, on very similar lines to industrial production in the West, also leads to the destruction of the environment in the countries where the capitalist class has been expropriated.

The consequences of this position are extremely grave. It is true that it is possible in this manner to derive the catastrophic ecological situation from the capitalist mode of production; but the more fundamental the categories, the more universal the result. The argument is irrefutable in an abstract sense, but it remains politically impotent. The statement that "capitalism is to blame" is correct in principle, but threatens to dwindle into an abstract negation of the existing order of things. Marxism is not a theory that exists in order to produce eternal verities; it is no good for Marxists to be right "in principle" when that means the end of the world.

Perhaps one has to remember that Marx represented *historical* materialism. From that it follows that the time factor cannot be eliminated from his theories. The delay in the coming of revolution in the overdeveloped capitalist lands is therefore not a matter of theoretical indifference. But that it was delayed does not in any way falsify the theory, for Marx certainly regarded the proletarian revolution as a necessary but not an automatic and inevitable consequence of capitalist development. He always maintained that there are alternatives in history and that the alternative facing the highly industrialized societies were long ago expressed in the formula "socialism or barbarism." In the face of the emerging ecological catastrophe this statement takes on a new meaning. The fight against the capitalist mode of production has become a race with time that mankind is in danger of losing. The tenacity with which that mode of production still asserts itself fifty years after the expropriation of the capitalist class in the Soviet Union indicates the kind of time dimensions we are discussing. It is an open question how far the destruction that it has wrought here on earth and continues to wreak is still reversible.

In this situation one must be relentless in critically examining certain elements in the Marxist tradition. First of all, one must examine to what extent one is dealing with original elements of Marxist thought or with later deformations of theory. Compared with the range of such questions the "preservation of the classics" seems a trifling matter. Catastrophes cannot be combated by quotations.

To begin with, one must critically examine the concept of material progress that plays so decisive a part in the Marxist tradition. It appears in any case to be redundant in that it is linked to the technical optimism of the nineteenth century. The revolutions of the twentieth century have led to victory in industrially underdeveloped countries and thereby falsified the idea that the socialist revolution was tied to a certain degree of "ripeness" and to "the development of the productive forces," or was actually the outcome of a kind of natural necessity. On the contrary, it has been demonstrated that "the development of the productive forces" is not a linear process to which political hopes can be attached:

> Until a few years ago most Marxists accepted the traditional view that the development of the productive forces was by its nature positive. They were persuaded that capitalism, in the course of its development, would provide a material base that would be taken over by a socialist society—one on which socialism could be constructed. The view was widely diffused that socialism would be more easily developed the higher the development of the productive forces. Productive forces like technology, science, human capabilities and knowledge, and a surplus of reified labor would considerably facilitate the transition to socialism.
>
> These ideas were somewhat mechanistically based on the Marxist thesis of the sharpening of the contradictions between the productive forces, on the one hand, and the relationships of production, on the other. But one can no longer assume that the productive forces are largely independent of the relationship of production and spontaneously clash with them. On the contrary, the developments of the last two decades lead one to the conclusion that the productive forces were formed by the capitalist productive relationships and so deeply stamped by them that any attempt to alter the productive relationships must fail if the nature of the productive forces—and not merely the way they are used—is not changed.[20]

Beyond a certain point, therefore, these productive forces reveal another aspect that till now was always concealed, and reveal themselves to be destructive forces, not only in the particular sense of arms manufacture and built-in obsolescence, but in a far wider sense. The industrial process, insofar as it depends on these deformed productive forces, threatens its very existence and the existence of human society. This development is damaging not only to the present but to the future as well, and with it—at least as far as our "Western" societies are concerned—to the utopian side of communism. If nature has been damaged to a certain, admittedly not easily determinable, degree and if that damage is irreversible, then the idea of a free society begins to lose its meaning. It seems completely absurd to speak in a short-term perspective, as Marcuse has done, of a "society of superabundance" or of the abolition of want. The "wealth" of the overdeveloped consumer societies of the West, insofar as it is not a mere mirage for the bulk of the population, is the result of a wave of plunder and pillage unparalleled in history; its victims are, on the one hand, the peoples of the Third World and, on the other, the men and women of the future. It is therefore a kind of wealth that produces unimaginable want.

The social and political thinking of the ecologists is marred by blindness and naïveté. If such a statement needs to be proven, the review of their thinking that follows will do so. Yet they have one advantage over the utopian thinking of the Left in the West, namely, the realization that any possible freedom and that every political theory and practice—including that of socialists—is confronted not with the problem of abundance, but with that of survival.

WHAT ECOLOGY PROPOSES

Most scientists who handle environmental problems are not visible to the general public. They are highly specialized experts, exclusively concerned with their carefully defined research fields. Their influence is usually exerted in the role of advisers. When doing basic research they tend to be paid from public fields; those who have a closer relationship with industry are predominantly experts whose results have immediate application. Most nonspecialists, however, aim to have a direct influence on the public. It is they who write alarmist articles that are published in magazines such as *Scientific American* or *Science*. They appear on television, organize congresses, and write the best-sellers that have shaped the picture of ecological destruction most of us imagine. Their ideas as to what should be done are reflected in the reforms promised by parties and governments. They are in this sense representative of something. What they say in public does not convey how valid their utterances are as scientific statements, yet it is important to analyze their proposals, for they indicate where the lines of scientific exploration and the dominant "bourgeois" ideology intersect.

The Americans Paul and Anne Ehrlich were among the founders of human ecology, and they still are two of its most influential spokespeople. In their handbook on ecology they summarize their proposals under the heading "A Positive Program," excerpts from which are extremely revealing:

2. Political pressure must be applied immediately to induce the United States government to assume its responsibility to halt the growth of the American population. Once growth is halted, the government should undertake to regulate the birthrate so that the population is reduced to an optimum size and maintained there. It is essential that a grass-roots political movement be generated to convince our legislators and the executive branch of the government that they must act rapidly. The program should be based on what politicians understand best— votes. Presidents, Congressmen, Senators, and other elected officials who do not deal effectively with the crisis must be defeated at the polls and more intelligent and responsible candidates elected.

3. A massive campaign must be launched to restore a quality environment in North America and to *de-develop the United States.* De-development means bringing our economic system (especially patterns of consumption) into line with the realities of ecology and the world resource situation. . . . Marxists claim that capitalism is intrinsically expansionist and wasteful, and that it automatically produces a monied ruling class. Can our economists prove them wrong? . . .

5. It is unfortunate that at the time of the greatest crisis the United States and the world has ever faced, many Americans, especially the young, have given up hope that the government can be modernized and changed in direction through functioning of the elective process. Their despair may have some foundation, but a partial attempt to institute a "new politics" very nearly succeeded in 1968. In addition many members of Congress and other government leaders, both Democrats and Republicans, are very much aware of the problems outlined in this book and are determined to do something about them. Others are joining their ranks as the dangers before us daily become more apparent. These people need public support in order to be effective. The world cannot, in its present critical state, be saved by merely tearing down old institutions, even if rational plans existed for constructing better ones from the ruins. We simply do not have the time. Either we will succeed by bending old institutions or we will succumb to disaster. Considering the potential rewards and consequences we see no choice but to make an effort to modernize the system. It may be necessary to organize a new political party with an ecological outlook and national and international orientation to provide an alternative to the present parties with their local and parochial interests. The environmental issue may well provide the basis for this.

6. Perhaps the major necessary ingredient that has been missing from a solution to the problems of both the United States and the rest of the world is a goal, a vision of the kind of Spaceship Earth that ought to be and the kind of crew that should man her.[21]

This is not the only case of a serious scientist presenting the public with an alarmist program of this kind. On the contrary, page upon page could be used to document similar ideas. They can be seen as a consensus of what modern ecology has to offer in the way of suggestions for social action. A collection of similar statements would only repeat itself; and we will therefore confine ourselves to one further piece of evidence. The following quotation is from a book by the Swede Gösta Ehrensvärd, a leading biochemist, in which he attempts a comprehensive diagnosis of the ecological situation. His therapeutic ideas can be summarized as follows:

> We are not *compelled* to pursue population growth, the consumption of energy, and unlimited exploitation of resources, to the point where famine and worldwide suffering will be the results. We are not *compelled* to watch developments and do nothing and to pursue our activities shortsightedly without developing a long-term view.

The catastrophe can be avoided, he says,

> if we take certain measures *now* on a global scale. These measures could stabilize the situation for the next few centuries and allow us to bring about, with as little friction as possible, the transition from today's hectically growing industrialized economy to the agricultural economy of the future. The following components of a crash program are intended to gain time for the necessary global restructuring of society on this earth.
>
> 1. Immediate introduction of worldwide rationing of all fossil fuels, above all of fluid resources of energy. Limitation of energy production to the 1970 level. Drastic restrictions on all traffic, insofar as it is propelled by fluid fuels, and is not needed for farming, forestry, and the long-distance transport of raw materials.
>
> 2. Immediate total rationing of electricity.
>
> 3. Immediate cessation of the production of purely luxury goods and other products not essential for survival, including every kind of armament.
>
> 4. Immediate food rationing in all industrial countries. Limitation of all food imports from the developing countries to a minimum. The main effort in terms of development policies throughout the world to be directed toward agriculture and forestry.
>
> 5. Immediate imposition of the duty to collect and recycle all discarded metal objects, and in particular to collect all scrap.
>
> 6. Top priority to be given to research on the development of energy from atomic fusion as well as to biological research in the fields of genetics, applied ecology, and wood chemistry.
>
> 7. Creation of an International Center to supervise and carry through action around the six points listed above. This Center to have the duty to keep the inhabitants of this earth constantly informed through the mass media of the level of energy and mineral reserves, the progress of research, and the demographic situation.[22]

A CRITIQUE OF THE ECOLOGICAL
CRASH PROGRAM

In their appeals to a world whose imminent decline they prophesy, the spokesmen of human ecology have developed a missionary style. They often employ the most dramatic strokes to paint a future so black that after reading their works one wonders how people can persist in giving birth to children or in drawing up pension schemes. Yet at the conclusion of their sermons, in which the inevitability of the end—of industrialization, of civilization, of man, of life on this planet—is convincingly described—if not proved—another way forward is presented. The ecologists end up by appealing to the rationality of their readers; if everyone would grasp what is at stake then—apparently—everything would not be lost. These sudden about-turns smack of conversion rhetoric. The horror of the predicted catastrophe contrasts sharply with the mildness of the admonition with which we are allowed to escape. This contrast is so obvious and so central that both sides of the argument undermine each other. At least one of them fails to convince, either the final exhortation, which addresses us in mild terms, or the analysis that is intended to alarm us. It is impossible not to feel that those warnings and threats, which present us with the consequences of our actions, are intended precisely to soften us up for the conversion that the anxious preacher wishes to obtain from us in the end; conversely, the confident final resolution should prevent us from taking too literally the dark picture they have painted, and from sinking into resignation. Every parish priest is aware of this noble form of verbal excess; and everyone listening can easily see through it. The result is (at best) a pleasurable *frisson.* Herein may lie the total inefficacy of widely distributed publications maintaining that the hour will soon come not only for man himself, but for all species. They are as ineffective as the typical Sunday sermon.

In its closest details, both the form and content of the Ehrlichs' argument are marked by the consciousness (or rather the unconsciousness) of the WASP, the white Protestant middle-class North American. This is especially obvious in the authors' social and political ideas: they are just as *unwilling* to consider any radical interference with the political system of the United States as they are *willing* to contemplate the other immense changes they spell out. The U.S. system is introduced into their calculations as a constant factor: it is introduced not as it is, but as it appears to be to the white member of the middle class, that is to say, in a form that has been transformed out of recognition by ideology. Class contradictions and class interests are completely denied: the parliamentary mechanism of the vote is unquestionably considered to be an effective method by means of which all conceivable conflicts can be resolved. It is merely a question of finding the right candidate and conducting the right campaigns, of writing letters and launching a few modest citizens' activities. At the most extreme, a new Congress will have to be set up. Imperialism does not exist. World peace will be reached through disarmament. The political process is posed in highly personalized terms: politics is the business of the politicians who are expected to carry the "responsibility." Similarly, economics is the business of the economists, whose task is to "draw up" a suitable economic system—this, at least, one has the right to ask of them. The term "Marxist" appears only once, as a scarecrow to drive recalcitrant readers into the authors' arms. All that this crude picture of political idiocy lacks are lofty ideas: the authors are not averse to make good the lack. What is needed is a "vision," since only relatively "idealistic programs" still offer the possibility of salvation. Since the need is so great, there will not be a lack of offers, and the academic advertising agency promptly comes up with the concept of "Spaceship Earth," in which the armaments industry and public relations join hands. The depoliticization of the ecological question is now complete. Its social components and consequences have been entirely eliminated.

Concrete demands can now cheerfully be made. There is no danger that they may be implemented with disagreeable consequences. A brake on population increase, de-development of the economy, or draconian rationing can now be presented as measures that, since they are offered in a spirit of enlightened, moral common sense, and are carried out in a peaceful, liberal manner, harm no interests or privileges and demand no changes in the social and economic system. Ehrensvärd presents the same demands in more trenchant, apparently radical terms—those of the coolly calculating scientist. Like the Ehrlichs, his arguments are so unpolitical as to be grotesque. Yet his sense of reality is strong enough for him to demand privileges for himself and his work—that is to say, he expects the highest priority for the undisturbed continuation of his research. One particular social interest, if a very restricted one, thereby finds expression: his own.

"Many of the suggestions," say the Ehrlichs, "will seem 'unrealistic,' and indeed this is how we view them."[23] The fact that not even the authors take their own "crash program" seriously at least makes it clear that we are not dealing with madmen. They reason why they seek refuge in absurdity is that their competence as scientists is limited precisely to the theoretical radius of the old ecology, that is to say, to a subordinate discipline of biology. They have extended their researches to human society, but they have not increased their knowledge in any way. It has escaped them that human existence remains incomprehensible if one totally disregards its social determinants; that this lack is damaging to all scientific utterances on our present and future; and that the range of these utterances is reduced whenever these scientists abandon the methodology of their particular discipline. It is restricted to the narrow horizons of their own class. The latter, which they erroneously regard as the silent majority is, in fact, a privileged and very vocal minority.

CONCLUSIONS: HYPOTHESES
CONCERNING A HYPOTHESIS

There is a great temptation to leave matters there and to interpret the forecast of a great ecological crisis as a maneuver intended to divert people from acute political controversy. There are even said to be parts of the Left that consider it a luxury to trouble themselves with problems of the future. To do that would be a declaration of bankruptcy; socialist thinking has from the beginning been oriented not toward the past but toward the future. Herein lay one of its real chances of success. For while the bourgeoisie is intent on the short-term interests of the accumulation of capital, there is no reason for the Left to exclude long-term aims and perspectives. As far as the competence of the ecologists is concerned, it would be a mistake to conclude that, because of their boundless ignorance on social matters, their statements are absolutely unfounded. Their methodological ineptitude certainly decreases the validity of their overall prognoses; but individual lines of argument, which they found predominantly on the causality of the natural sciences, are still usable. To demonstrate that they have not been thought through in the area of social causes and effects is not to refute them.

The ideologies of the ruling class do not reproduce mere falsifications. Even in their instrumental form they still contain experiences that are real insofar as they are never optimistic. They promise the twilight of the gods, global catastrophe, and a last judgment; but these announcements are not seen to be connected with the identification and short-term satisfactions that form part of their content.[24]

All this applies admirably to the central "ecological hypothesis" according to which if the present process of industrialization continues naturally it will in the foreseeable future have catastrophic results. The central core of this hypothesis can neither be proved nor refuted by political discussion. What it says is of such importance, however, that what one is faced with is a calculation like Pascal's wager. So long as the hypothesis is not unequivocally refuted, it will be heuristically necessary to base any thinking about the future on what it has to say. Only if one behaves "as if" the ecological hypothesis was valid can one test its social validity—a task that has scarcely been attempted up to now and of which ecology itself is clearly incapable. The following reflections are merely some first steps along this path. They are, in other words, hypotheses based on other hypotheses.

A general social definition of the ecological problem would have to start with the mode of production. Everywhere the capitalistic mode of production obtains totally or predominantly—that is to say, where the products of human labor take the form of commodities—increasing social want is created alongside increasing social wealth. This want assumes different forms in the course of historical development. In the phase of primitive accumulation it expresses itself in direct impoverishment caused by extensive exploitation, extension of working hours, and the lowering of real wages. During cyclical crises, the wealth that has been produced by labor is simply destroyed—grain is thrown into the sea and so on. With the growth of the productive powers the destructive energies of the system also increase. Further want is generated by world wars and armaments production. In a later phase of capitalistic development this destructive potential acquires a new quality. It threatens all the natural bases of human life. This has the result that want appears to be a socially produced natural force. This return of general shortages forms the core of the "ecological crisis." It is not, however, a relapse into conditions and circumstances from the historical past because the want does not in any sense abolish the prevailing wealth. Both are present at one and the same time; the contradiction between them becomes ever sharper and takes on increasingly insane forms.

So long as the capitalist mode of production obtains—that is to say, not merely the capitalist property relationships—the trend can at best be reversed in detail but not in its totality. The crisis will naturally set in motion many processes of adaptation and learning. Technological attempts to level out its symptoms in the sense of achieving a homeostasis have already gone beyond the experimental stage. The more critical the situation becomes, the more desperate will be the attempts undertaken in this direction. They will include abolition of the car, construction of means of mass transport, erection of plants for the filtration and desalination of seawater, the opening up of new sources of energy, synthetic production of raw materials, the development of more intensive agricultural techniques, and so on. But each of these steps will cause new critical problems; these are stopgap techniques that do not touch the roots of the problem. The political consequences are clear enough. The costs of living accommodation and space for recreation, of clean air and water, of energy and raw materials will increase explosively, as will the cost of recycling scarce resources. The "invisible" social costs of capitalist commodity production are rising immeasurably and are being passed on in prices and taxes to the dependent masses to such a degree that any equalization through controlling wages is no longer possible. There is no question, needless to say, of a "just" distribution of shortages within the framework of Western class society: the rationing of want is carried out through prices, and if necessary through gray or black markets, by means of corruption and the sale of privileges. The subjective value of privileged class positions increases enormously. The physiological and psychic consequences of the environmental crisis, the lowered expectation of life, and the direct threat from local catastrophes can lead to a situation where class can determine the life or death of an individual by deciding such factors as the availability of means of escape, second houses, or advanced medical treatment.

The speed with which these possibilities will enter the consciousness of the masses cannot be predicted. It will depend on the point in time at which the creeping nature of the ecological crisis becomes apparent in spectacular individual cases. Even dramatic phenomena such as have principally appeared in Japan—the radioactive poisoning of fishermen, illnesses caused by mercury and cadmium—have not yet led to a more powerful mobilization of the masses because other consequences of the contamination have become apparent only months or years later. But once, at any point in the chain of events, many people are killed, the indifference with which the prognoses of the ecologists are met today will turn into panic reaction and even into ecological rebellions.

There will, of course, be organizational initiatives and political consequences at an even earlier stage. The ecological movement in the United States, with its tendency to flee from the towns and industry, is an indication of what will come, as are the citizen's campaigns that are spreading apace. The limitations that beset most of these groups are not fortuitous; their activity is usually aimed at removing a particular problem. There is no other alternative, for they can only crystallize around particular interests. A typical campaign will, for example, attempt to prevent the siting of an oil refinery in a particular district. That does not lead, if the agitation is successful, to the project being canceled or to a revision of the policy on energy; instead, the refinery is merely built where the resistance of those affected is less strongly expressed. In no case does the campaign lead to a reduction of energy consumption. An appeal on these grounds would have no sense. It would fall back on the abstract, empty formulae that make up the "crash programs" of the ecologists.

The knot of the ecological crisis cannot be cut with a paper knife. The crisis is inseparable from the conditions of existence systematically determined by the mode of production. That is why moral appeals to the people of the "rich" lands to lower their standard of living are totally absurd. They are not only useless but cynical. To ask the individual wage earner to differentiate between his "real" and his "artificial" needs is to mistake his real situation. Both are so closely connected that they constitute a relationship that is subjectively and objectively indivisible. Hunger for commodities, in all its blindness, is a product of the production of commodities, which could only be suppressed by force. We must reckon with the likelihood that bourgeois policy will systematically exploit the resulting mystifications—increasingly so, as the ecological crisis takes on more threatening forms. To achieve this, it only needs demagogically to take up the proposals of the ecologists and give them political circulation. The appeal to the common good, which demands sacrifice and obedience, will be taken up by these movements together with a reactionary populism determined to defend capitalism with anticapitalist phrases.

In reality, capitalism's policy on the environment, raw materials, and population will put an end to the last liberal illusions. That policy cannot even be conceived without increasing repression and regimentation. Fascism has already demonstrated its capabilities as a savior in extreme crisis situations and as the administrator of poverty. In an atmosphere of panic and uncontrollable emotions—that is to say, in the event of an ecological catastrophe that is directly perceptible on a mass scale—the ruling class will not hesitate to have recourse to such solutions. The ability of the masses to see the connection between the mode of production and the crisis in such a situation and to react offensively cannot be assumed. It depends on the degree of politicization and organization achieved by then. But it would be facile to count on such a development. It is more probable that what has been called "internal imperialism" will increase. What Negt and Kluge have observed in another connection is also relevant to the contradiction between social wealth and social poverty, which is apparent in the ecological crisis: "Colonialization of the consciousness or civil war are the extreme forms in which these contradictions find public expression. What precedes this collision, or is a consequence of it, is the division of individuals or of social groups into qualities which are organized against each other."[25]

In this situation, external imperialism will also regress to historically earlier forms—but with an enormously increased destructive potential. If the "peaceful" methods of modern exploitation fail, and the formula for coexistence under pressure of scarcity snaps, then presumably there will be new predations, competitive wars, and wars over raw materials. The strategic importance of the Third World, above all of those lands that export oil and nonferrous metals, will increase, and with it their consciousness that the metropolitan lands depend on them. The "siege" of the metropolises by the village—a concept that appeared premature in the 1950s—will acquire quite new topicality. It has already been unmistakably heralded by the policy of a number of oil-producing countries. Imperialism will do everything to incite the population of the industrialized countries against such apparent external enemies whose policy will be presented as a direct threat to their standard of living, and to their very survival, in order to win their assent to military operations.

Talk in global terms about Spaceship Earth tells us almost nothing about real perspectives and the chances of survival. There are certainly ecological factors whose effect is global; among these are macroclimatic changes, pollution by radioactive elements, and poisons in the atmosphere and oceans. As the example of China shows, it is not these overall factors that are decisive, but the social variables. The destruction of mankind cannot be considered a purely natural process. But it will not be averted by the preachings of scientists, who only reveal their own helplessness and blindness the moment they overstep the narrow limits of their own special areas of competence. "The human essence of nature first exists only for social man; for only here does nature exist as the foundation of his own human existence. Only here has what is to him his natural existence become his human existence, and nature become man for him. Thus society is the unity of being of man with nature—the true resurrection of nature—the naturalism of man and the humanism of nature both brought to fulfillment."[26]

If ecology's hypotheses are valid, then capitalist societies have probably thrown away the chance of realizing Marx's project for the reconciliation of man and nature. The productive forces that bourgeois society has unleashed have been caught up with and overtaken by the destructive powers released at the same time. The highly industrialized countries of the West will not be alone in paying the price for the revolution that never happened. The fight against want is an inheritance they leave to all mankind, even in those areas where mankind survives the catastrophe. Socialism, which was once a promise of liberation, has become a question of survival. If the ecological equilibrium is broken, then the rule of freedom will be further off than ever.

NOTES

1. Murray Bookchin, *Ecology and Revolutionary Thought* (New York: 1970), 11. Bookchin argues that to ask an ecologist exactly when the ecological catastrophe will occur is like asking a psychiatrist to predict exactly when psychological pressure will so affect a neurotic that communication with him will become impossible.

2. Report from the Poor Law Commissioners to the Home Department, *An Inquiry into the Sanitary Conditions of the Labouring Population of Great Britain* (London: 1842), 68. Quoted in James Ridgeway, *The Politics of Ecology* (New York: Dutton, 1971).

3. Examples of this are not lacking in the ecology movement. In France there is an organization for environmental protection that has an extremely right-wing orientation. The president of these "ecofascists" is none other than General Massu, the man responsible for the French use of torture in the Algerian war.

4. Rote Reihe, *Profitschmutz und Umweltschmutz* (Heidelberg: 1973), 1:5.

5. Karl Marx, *Capital* (Moscow: 1961), 1:510fn.

6. Ridgeway (*Politics of Ecology,* 22–25) sees Chadwick as an archbetypical utilitarian bureaucrat, whose function was to secure the interests of capital by achieving peace and order among the poor. Better sanitation would produce a healthier and longer living work force, sanitary housing would raise workers' morale, and so on.

7. Ridgeway (*Politics of Ecology,* 15ff.) shows that over 150 years ago the Benthamites had evolved a theory of protecting the environment to promote production. As he also points out, the measures taken in the advanced capitalist United States in the late 1960s fail to reach the standards of water and air cleanliness proposed by the utilitarians.

8. *Der Spiegel,* January 8, 1973, p. 38.

9. Ridgeway (*Politics of Ecology,* 207–211) analyses the "eco–industrial complex," that is, the growing role played by business in promoting ecological campaigns, such as Earth Day, and the liasons between business, politicians, local governments, and "citizen campaigns."

10. For illustration of the "eco-industrial complex" in West Germany, see Reihe, *Profitschmutz,* p. 14, and the pamphlet *Ohne uns hein Umweltschutz.*

11. "Primera Conferencia de Solidaridad de los Pueblos de América Latina," in *América Latina: Demografía, población indígena y salud* (Havana: 1968), 2:15ff.

12. Ibid., 55–57.

13. Claus Koch, "Mystífikationen der 'Wachstumskrise.' Zum Bericht des Club of Rome," *Merkur, 297,* January, 1973, 82.

14. Giorgio Nebbia, preface to his *La Morte ecologica,* (Bari, Italy: 1972), xvff.

15. *The Limits of Growth: Report of the Club of Rome on the State of Mankind* (London: 1972), 13.

16. Gerhard Kade, *Kapitalismus und "Umweltkatastrophe"* (duplicated manuscript, 1973).

17. Rossana Rossanda, "Die sozialistischen Länder: Ein Dilemma des westeuropäischen Linken," *Kursbuch, 30,* 1973, 26.

18. Cf. "Marx und die Oekologie," *Kursbuch, 33,* 1973, 175–187.

19. Rossanda, "Die sozialistischen Länder," 30.

20. André Gorz, "Technique, techniciens et lutte de classes," *Les Temps Modernes, 30*(12), August–September, 1971, 141.

21. Anne H. Ehrlich and Paul R. Ehrlich, *Population, Resources, Environment* (San Francisco: 1970), 322–324.

22. Gösta Ehrensvärd, *Före-efter, En Diagnos* (Stockholm: 1971), 105–107.

23. Ehrlich and Ehrlich, *Population,* 322.

24. *Öffentlichkeit und Erfahrung. Zur Organisationsanalyse von bürgerlicher und proletarischer Offentlichkeit* (Frankfurt: 1972), 242.

25. Ibid., 283ff.

26. Karl Marx, *Economic and Philosophical Manuscripts of 1844,* ed. D. Struik (London: 1970), 137.

Chapter 2

MARXISM AND ECOLOGY
More Benedictine Than Franciscan

JEAN-GUY VAILLANCOURT

INTRODUCTION

Various writers interested in ecology and Marxism have attempted to identify the connections—whether of similarity or difference—between these two important currents of thought, both of which appeared in Europe around the middle of the last century. Such a task, needless to say, is quite difficult. Indeed, it is not surprising that some writers, located within as well as outside of Marxism,[1] have held that Marxism and ecology are diametrically opposed, or that ecology is a sort of substitute for Marxism[2]; while others have argued for a close affinity between the two,[3] even to the point of saying that the basic ideas of ecology were anticipated in Marx and Engels's work.[4]

Certain ambiguities in this regard may result from the ambivalence of Marxism toward man's domination of nature, but in my opinion the greatest problem arises from the fact that we find it difficult to define what we mean by "Marxism" and what we mean by "ecology." In truth, there are many ways of being a Marxist or of being an ecologist. This chapter will be concerned with the beginnings of the science of ecology as well as with present-day ecologists. To classify the latter, it suffices to recall here the three major types of ecologists who have successively taken a leading role in the ecological movement over the past century: (1) preservationists and conservationists, (2) environmentalists, and (3) political ecologists.[5]

Ecology and Marxism have undergone considerable change over the past hundred years, and each of these approaches has engendered numerous tendencies and perspectives, sometimes quite at odds with each other. I shall not, therefore, attempt to compare all the different types of Marxism and ecology. I will limit myself in this respect to the initial period of theoretical development to see if preoccupations that could be considered ecological or antiecological can be found in the writings of Marx (1818–1883) and Engels (1820–1895). These two founders of Marxism left in their writings—including their correspondence—numerous references to the man–nature relationship, urban and industrial pollution, the depletion of resources, and also to the ideas of the forerunners and even the founders of ecology. Thus, before saying that Marxism is, in general, in agreement with or opposed to ecology, it is essential to see how Marx and Engels reacted to ecological science and problems in their day by quoting from their original texts. So much is said about Marx and Engels that it seems better to let them to speak for themselves. I think it is because they have taken Marx's own ideas as their starting point, and because they have distinguished between different approaches to ecology and the environment, that authors like Syer and Clow have achieved such balanced positions concerning the relations between Marxism and ecology.[6]

In the text that follows, I will examine chronologically certain writings of Marx and Engels in order to identify the major pro- or antiecological references they contain.[7] Then, in conclusion, I will try to answer this question: Were Marx and Engels precursors of ecology? Were they, in short, sensitive to the ecological preoccupations that first appeared a century ago?

THE EARLY WRITINGS OF MARX AND ENGELS

From the doctoral thesis of Marx, presented at the University of Iena in 1841, on the philosophy of nature of Democritus and Epicurus,[8] to Engels's posthumous and uncompleted work on the dialectics of nature, the works of the two founders of Marxism contain not only numerous references to many of the forerunners and founders of different branches of ecology, but also ideas that still remain of great importance to ecologists today. Democritus and Epicurus were Greek philosophers of nature who lived more than a century after Parmenides and Heraclitus, and who can be considered along with the latter, and like Hippocrates, Aristotle, and Theophrastus, as the forerunners of the sciences of nature and even of scientific ecology. These two philosophers taught that everything that exists is made up of empty space and of an infinity of atoms in perpetual movement, combined in different ways. It was Epicurus who said that nothing is created, nothing is destroyed. The four fundamental laws of ecology, according to Commoner[9]—that everything is linked to all else, that everything must go somewhere, that nature is always right, and that nothing is free—owe a great deal to these two Greek philosophers, who were the first to establish the foundation for a serious analysis of nature.

In the *Economic and Philosophical Manuscripts of 1844,* the young Marx expresses for the first time, in a fairly elaborate way, his ideas about what communism should make of the relationship between man and nature. The emphasis is placed on materialism and naturalism, and on the compatibility of these with humanism.[10] This work also contains a bitter condemnation of capitalism's terrible effects on the quality of life of urban workers.[11]

In his "Introduction to the Critique of Political Economy," published in the *Annales Franco-Allemandes,* also in 1844, Engels virulently attacked the private appropriation of land by capital.[12] Engels was also concerned, at the outset of his intellectual career, with environmental problems. In this same text, he expressed his optimism concerning the productive forces and science: "The productive forces at humanity's disposal are limitless; the yield from the earth can increase indefinitely through the application of capital, work and science."[13] Malthusianism is presented here as "this infamous and vile doctrine, this abominable blasphemy against man and nature."[14] Nevertheless, Engels does not completely rule out the possibility that Malthus might be right, although he believes that, even in that case, communism would still be necessary.[15] Lastly, having paid homage to Davey and Liebig, Engels reaffirmed his optimism toward science: "What progress does agriculture in this century not owe chemistry, indeed, uniquely to two people—Sir Humphrey Davey and Justus Liebig? . . . But then, what is not possible for science?"[16]

In the following year, Engels published an important work which placed him immediately among the forerunners of human ecology. This book, on the situation of the working class in England,[17] is a description of working-class conditions which in many ways resembles such great social investigations of the time as the one written by Villermé in France.[18]

This investigation by Engels falls into the tradition of the British "social surveys" and the French monographs that represented, before their time, the first studies in urban ecology. A major concern is the harmful effects of capitalism on workers' health. Among other things, Engels points to the fact that the atmosphere in London and Manchester contains much more carbon dioxide and much less oxygen than the atmosphere in the countryside. This work contains interesting descriptions of the ecological crisis experienced by miners, farm laborers, and industrial workers in their working and living environments.[19] In his conclusion, Engels makes another bitter critique of Malthus's theories, which legitimated the harsh treatment meted out to the poor by the English ruling class. This is a theme to which Engels and Marx will often return in their later works. Indeed, in their numerous criticisms of Malthus, they anticipate some of the objections later raised by (ecological or nonecological) radicals against Malthus and his disciples.

In the following texts, the first of which is an extract from his pamphlet on the Jewish question, the other from the *Grundrisse* of 1857–1858, Marx displays his fundamental ambiguity vis-à-vis the thorny question of man's subjugation of nature. On the one hand, he shows how capitalism holds nature in contempt: "The way in which nature is perceived under the law of private property is one of utter scorn, a debasement of nature, in effect."[20]

On the other hand, when he describes the communism of the future, his framework remains predicated on the subjugation of nature. For him, communism "will be the fully developed domination of man over natural forces, over nature in the strict sense, as well as over his own nature."[21] Later, in the chapter on capital, Marx takes issue with what "this monkey of a Malthus" has said of the relationship between human population and food.[22]

In the Marx–Engels correspondence, especially after 1859, and in Engels's *Anti-Dühring* and the *Dialectics of Nature*, much is said of Malthus and Darwin, as well as of Haeckel and other disciples of Darwin. It is impossible to cite here all the pertinent passages, but I will at least mention a few. In a letter to Marx dated December 12, 1859, Engels writes: "This Darwin I am now reading, is quite sensational. . . . No one has ever made an attempt on such a scale to demonstrate the existence of an historical dynamic in nature, at least never with such success."[23]

A year later, on December 19, 1860, it is Marx who writes to Engels about Darwin's book on natural selection: "Despite the inauspicious style of argument—so typically English—it is in this book that the historico-natural foundations of our theory can be found."[24] And less than a month after that, in a letter to Lassalle of January 16, 1861, Marx returns to some of Engels's ideas.[25] A year and a half later, he speaks to Engels again about Darwin, this time referring to Malthus:

> What amuses me about Darwin, which I have been looking at again, is his statement that he is applying the theory of "Malthus" to plants and animals, as if the cleverness of Mr. Malthus did not consist precisely in this, that his theory is applied not to plants and animals, but, along with the law of geometric progression, only to men—in direct opposition to plants and animals. It is remarkable to see how Darwin recognizes in

the animals and plants his own English society, with its division of labor, its competition, its opening of new markets, its "inventions," and its Malthusian "struggle for life." It is the *bellum omnium contra omnes* [the war of all against all] of Hobbes, which reminds one of Hegel in the *Phenomenology,* where civil society appears as "the fleshly realm of the Spirit," whereas with Darwin, it is the animal realm that takes the form of civil society.[26]

These ideas on the relationship between Malthus and Darwin are taken up by Engels in his letter of March 20, 1865, to Lange, and in that to Kugelman of June 27, 1870.[27]

CAPITAL

It is in *Capital* that the ideas of Marx concerning man–nature relations, industrial pollution, and agriculture's exhaustion of the soil are the most completely and clearly expressed. I cite here some of the most pertinent references. In volume 1, Marx speaks of the relative importance of work and of land as sources of wealth.[28] A little further on in this same volume, Marx refers to the parallels between the exploitation of workers by capital and the exploitation of the soil by the famished husbandman.[29] We should recall once more the well-known passage on manufacturing in which Marx denounces industrial pollution and speaks about the health and safety of the workers,[30] since these are two of the major preoccupations of contemporary ecologists. Another passage often quoted by present-day ecologists is the paragraph on large-scale industry and agriculture in which Marx presents a brilliant analysis of the connection between the exploitation of land and the exploitation of labor.[31] Lastly, let me mention as well a note in which Marx refers to Liebig, who has "clearly brought to light the negative side of modern agriculture, from a scientific point of view."[32]

In April and May of 1863, Engels writes in his letters to Marx about the forerunners of Darwin: Lyell, Schmerling, and Barthes, as well as Thomas Huxley. A few years later, in 1866 and 1868, Marx launches a minipolemic with Engels about the work of Tremaux, who claims to be a geological determinist. He also mentions Fraas, who shows that climate and flora have evolved in the period of known history: "He [Fraas] is a Darwinist before Darwin, and demonstrates the development of the species themselves in the historical period."[33]

In his letter of February 15, 1869, to his daughter Laura and to Lafargue, Marx summarizes his opinion of Darwin and Darwinism:

Darwin was led, from the struggle for life in English society—the war of all against all, *bellum omnium contra omnes*—to discover that the struggle for life was the dominant law in "animal" and plant life as well. As for the Darwinian movement, however, it sees in that a compelling reason for human society to never emancipate itself from its animal character.[34]

Marx had great esteem for Darwin. Thus he sent him a complimentary copy of volume 1 of *Capital* with these words: "Mr. Charles Darwin. On the part of his sincere admirer/Karl Marx. . . . London. June 16, 1873, Modena Villas, Maitland Park." Darwin thanked him in a polite, reserved letter written on the first of October of that year.[35]

Some years later, however, Engels and Marx began to distance themselves from Darwin and especially from his conservative disciples. In a letter to Piotr Lavrov in November 1875, Engels briefly suggests how he was planning to attack such bourgeois Darwinists.[36] The following year, Marx complains to Engels that Darwin has participated in a national conference on the Eastern question engineered by the Liberal Party: "Charles Darwin, too, alas, has added his name to this ridiculous gathering."[37]

The admiration that Marx and Engels had for Darwin and his disciples waned considerably in the course of the 1870s. This cooling is evident in Engels's letter of July 19, 1878, to Schmidt, a Darwinist zoologist.[38] Yet this did not hinder Engels from speaking admiringly of Darwin during Marx's funeral: "Charles Darwin discovered the law of organic natural evolution on our planet. Marx is the one who discovered the fundamental and constitutive law which determines the course of the evolution of human history."[39]

In volume 2 of *Capital* (which Engels published in 1885, two years after the death of its author), Marx describes the plundering of the forests under capitalism in very modern terms, such as one might easily encounter today in the writings of political ecologists.[40] In volume 3, which only appeared in 1895, the year of Engels's death, Marx speaks about the pollution of the Thames River by sewage waste, which could have been usefully recycled for agriculture, and about the conservation of raw materials. Here again, Marx appears as a precursor.[41] Toward the end of this third volume, inspired by Liebig, Marx attacks once more capitalist agriculture, which exhausts the soil and endangers the future of the human race.[42] In *Theories of Surplus Value,* considered by many to be the fourth volume of *Capital,* Marx returns to his attack against Malthus, accusing him of having plagiarized not only Anderson but also Ricardo, Townsend, Steward, Wallace, Herbert, and others, and of being a brazen sycophant of the ruling classes as well as a cynical brute in his attitude toward the oppressed. Discussing the relationship between Malthus and Darwin, he writes:

> In his remarkable essay, Darwin fails to realize that, in attributing geometric increase to the plant and animal kingdoms, he is deviating from and changing Malthus' theory. Malthus' theory consists precisely

in contrasting the geometric increase of men, established by Wallace, with his imagined arithmetic progression of animals and plants. In the work of Darwin—as regards the extinction of species, for instance—we find the refutation of the material and historical character of Malthus' theory, not only in its fundamental principle, but even in its details.[43]

A bit further in the same essay, Marx sums up his opinion of Malthus's book *On Population* as follows: "Malthus' work . . . is a pamphlet directed against the French Revolution and the ideas of reform which were springing up at that time in England (Godwin, etc.); it served to justify the misery of the working class. His theory is a plagiarism of Townsend, etc."[44]

In my opinion, it is in *Capital* that one can find the most important passages concerning the ecological damage that men have inflicted on the natural environment, on the urban milieu, and on the working class. The extracts to which I have referred should by themselves suffice to show that the relationship between Marxism and ecology is not as antagonistic as some ecologists and some Marxists would try to make us believe, and that we can consider Marx as one of the precursors of human ecology.

ANTI-DÜHRING AND THE *DIALECTICS OF NATURE*

Let us now refer briefly to two later works of Engels that are more directly concerned with nature and ecology, *Anti-Dühring* and the *Dialectics of Nature*. In the preface to the former, Engels distinguished the kind of science of nature he advocates, and that of the philosophers of nature who had preceded him: "The philosophers of nature are to the consciously dialectical science of nature what the utopians are to modern Communism."[45] Once again, he returns to the Marxist critique of Darwin's and Malthus's ideas. This time around, however, Darwin is castigated as much as Malthus is.[46] One is far from the encomiums of the early 1860s. In another note, a bit further on, Engels speaks favorably about Haeckel's scientific ideas.[47]

Haeckel was not only the person who coined the word ecology, he was also the one who formulated the fundamental biogenetic law that ontogenesis is a short recapitulation of phylogenesis, that is, that the evolution of the individual is a brief recapitulation of the species' evolution as a whole. Here, then, was a scientist of considerable prestige whom Engels intended to use against the pseudoscientific and pseudosocialist Dühring. Engels was delighted that Haeckel made himself the defender of the freedom of scientific education against Virchow, but to the extent that Haeckel himself became antisocialist, to that same extent Engels changed his stand, eventually bitterly criticizing him. Their differences widened later on, especially after Haeckel managed to arrange that the University of Iena's first (honorary) doctorate in phylogenesis would be conferred on Bismarck, the chancellor.[48] He attracted Engels's wrath when, like Virchow, he ranged himself against socialism while attempting to show its lack of affinity with Darwinism. In a letter to Lavrov dated August 10, 1878, Engels wrote: "You will have seen that the German Darwinians, in responding to Virchow's call, have decidedly taken sides against socialism. Haeckel, whose brochure I have just received, limits himself to speaking in general terms of *verruckten Lehren des Sozialismus* [mad socialist doctrines], but Mr. Oscar Schmidt of Strasbourg will crush us *con amore* [with love] at the *Naturforscherversammlung* [the Congress of Naturalists] in Kassel."[49] Finally, Engels returns to some ideas enunciated earlier by Marx and himself on urban pollution.[50]

Engels's last work, the *Dialectics of Nature*, remained uncompleted. At its very beginning he paid homage to Kant, who introduced the notion of becoming into the study of nature, and to Lyell, who introduced reason into geology. Lamarck and especially Darwin are presented as precursors in the field of biological research. In the "Introduction" he compared animal societies and human societies, drawing on Darwin's ideas.[51] In the crucial chapter on "The Role of Work in Transforming Ape into Man," Engels provides us with a veritable lesson in general ecology. It is worth quoting in extenso:

> As we have already indicated, animals modify the natural environment as do men, although to a lesser degree, and, as we have seen, the modifications which they have effected in their surroundings react in turn to transform the actors themselves. For nothing happens in isolation in nature. Every phenomenon reacts on the other, and inversely, and it is usually because they forget this dynamic and this reciprocal action that your scholars are prevented from understanding the simplest things. We have seen how goats impeded the reforestation of Greece. . . . The animal destroys the vegetation without knowing what it is doing. Man destroys it to plant grain on the available land, or to plant trees or vines which he knows will yield him a harvest several times greater than what he has sown, transfers the useful plants and domestic animals from one country to another, and he thus modifies the flora and fauna of whole continents. There is more. Thanks to artificial selection, the agency of man transforms plants and animals to the point that they can no longer be recognized. . . . In brief, the animal only uses the external environment, modifying it simply by its presence; by the changes which he brings, man makes it serve his purposes. He conquers it.[52]

But he is very much aware that man's domination of nature can entail disastrous consequences:

Let us not, however, flatter ourselves too much as to our victories over nature. For each one of these, it takes its revenge. Initially, each victory certainly has the consequences which we anticipated, but at a second or third remove, it has effects that are so different, so unforeseen, that too often the initial achievement is destroyed.[53]

And he ends this chapter with a classic attack on the capitalist system, which only considers immediate profits without any consideration of the environmental and social effects of its actions.[54] In the chapter on the "Dialectic," Engels returns to the problem of the connection between man and nature, and he expresses in a clear manner the Marxist position on this subject:

That is why, in arguing that it is only nature that acts on man, that it is only natural conditions which everywhere shape his historical development, the naturalist conception of history—as it is manifested more or less in the works of Draper and some other scholars—is too unilateral, forgetting that man also reacts on nature, transforms it, creating new conditions of existence for himself. From the "nature" of Germany at the time when the Germans were settling there, there remains devilishly few things. The topsoil, the climate, the vegetation, the fauna and the men themselves have infinitely changed, and all this because of human activity, whereas the transformations which occurred during this time in the German environment, without any human involvement, are insignificant.[55]

In the chapter on the "Elements of the History of Science," Engels discusses the conception of nature of the scholars of antiquity—Aristotle, the Eleatic School, Democritus, and Epicurus—and mentions Lyell, Lamarck, and Darwin as having made great breakthroughs in the sciences. He criticizes Haeckel quite severely,[56] something that does not prevent him from citing his works several times in the chapter on "Biology," or from making favorable comparisons between his *Generale Morphologie des Organismen*—in which the word ecology is defined—and the work of Malthus and Darwin.[57] To summarize, Engels is aware of the writings of the precursors of ecology and of the first ecologists, and he makes extensive use of them in this work, attempting to show that the dialectic exists even in nature. Thus he too can be considered, in the same way as Marx, as a forerunner of human ecology.

CONCLUSION

Although they never employed the term "ecology," this science having begun to develop only toward the end of their lives, Marx and Engels, on the basis of what we have seen of their writings, can without doubt be considered as forerunners of human and social ecology, and environmental sociology, on the one hand, and of political ecology, especially of that tendency called "ecosocialist," on the other hand. As for the evolutionary ecology of plants and animals, which provides the basis for more recent developments at the human and social level, Marx, and especially Engels, not only recognized its emergence but were inspired by it in their own writings. The forerunners of biological ecology and of human ecology, as we have seen, were often cited in their writings from 1859 on.

Marx and Engels were especially sensitive to the interdependence of man and nature. Their materialism led them to highlight the importance of determination by material factors, such as space, natural resources, technology, and the like, but this is somewhat qualified and counterbalanced by the importance they give to economic and social realities and to dialectics.[58]

In reality, they waver back and forth between an anthropocentric framework, on the one hand, and a naturalist perspective that grows stronger especially after 1860, on the other—that is, from the time they discovered Darwin's ideas. As rebellious disciples of Hegel and Feuerbach, they were torn between their materialism, which sensitized them to the importance of the natural environment and the productive forces, and their humanism, which made them highlight socioeconomic determinants, historical change, and the dialectic. One can, in fact, discern a small difference between Marx and Engels on this point. For Marx, the dialectic is situated more within science, within the human context—that is, than within nature itself—while for Engels, especially in his later works, the dialectic is situated in the very heart of matter, independent of man. Marx attempted to maintain an equilibrium between materialism and humanism, while Engels seemed to want to return to a kind of naturalism not dissimilar to that of Feuerbach. In *Anti-Dühring* and the *Dialectics of Nature*, Engels moved from historical materialism—which he and Marx had earlier presented as a middle path between idealism and a mechanical, vulgar materialism—to a dialectical materialism in which emphasis is given more to nature than to man. But it should be said in his defense that he did this with the approval of Marx himself. This stance, which was frankly materialist, not to say naturalist, and which was ever more strongly emphasized by the founders of Marxism, is much more a philosophical option than a scientific one. Ecology is very careful not to adopt such a diffuse and all-encompassing *Weltanschauung*, preferring, rather, to remain at the scientific and empirical level (and at the political level, where political ecology is concerned). By extending the dialectic to matter, Engels, and to some extent Marx, endow it with spiritual qualities, indeed, with characteristics that are essentially divine. There is something messianic about the way in which they proclaim that communism will reconcile man to nature and man to his fellow man. In Marxism, the exploitation and oppression of the proletariat, the class chosen by history, leads to the end (or the beginning) of history, somewhat as in Christianity the sufferings and the sacrifice of God incarnate lead to the salvation of the world, nature included, and even to the divinization of creation. Ecology and political ecology do not have the sort of religious thrust that characterizes Marxism.

What is even more relevant to ecology, in the writings of Marx and Engels, is the critique of the social and environmental costs of capitalist production, and the concern for the quality of soils and for workers' health and their quality of life. There is even attention given to the beauty of nature, and to the conservation of resources, as well as to the protection of fauna and flora. For example, Marx notes approvingly the fact that Munzer, leader of the peasant and religious revolt, had protested against the private ownership of fish, birds, and plants. Moreover, Marx and Engels are very much "ecologists" in the way they link sociopolitical phenomena and the environment—for instance, in their blunt treatment of the human and natural destruction wreaked by capitalism, and their excoriation of the greed of private entrepreneurs, who plunder not only the forests but also the soil and subsoil. Despite certain progressive aspects, capitalism, in their view, both dehumanizes man and perverts the natural world. It is a parasitic and cannibalistic system, sucking the life from man and nature alike.

The human mastery of nature that Marx and Engels envision would entail not plunder or reckless exploitation, but a judicious management of the sort advocated by Francis Bacon, who says that man can and should modify nature, but only while remaining within the rules that it lays down, and that nature must be obeyed if one wishes to give it commands. They favor active and planned intervention as far as nature is concerned. In this regard, to employ the distinction made by René Dubos between the two types of contemporary ecologists, they are nearer the activist and meliorist "Benedictine" perspective than to the more contemplative and passive "Franciscan" approach.

For Marx and Engels, indeed, nature must be put to the service of man. They wanted to organize and develop production to satisfy human needs, while at the same time conserving the regenerating capacity of nature. The response of Marx and Engels to Malthus—that prophet of doom and spokesman of the landlords, state functionaries, and the middle class—was that there is no natural law that mandates a geometric increase in population and an arithmetic growth of food supply. According to them, land, which was still only lightly populated and developed in the middle of the last century, could sustain a growing population, provided that agricultural and industrial production would be organized in a rational and scientific fashion, under working-class control. They anticipated the damage that capitalism could inflict on mankind and on the planet, but they nonetheless remained fascinated with this system that had "saved a good part of the population from the imbecility of rural life," as they wrote in the *Communist Manifesto.*

Marx and Engels remained optimists as regards technological and economic progress, the abundance of resources, and the possibilities for human population growth—provided that these occur within a socialist framework and in the best interest of the working classes. They at once rejected both the restless greed of capitalism and the bucolic passivity of some of the utopian socialists. Their position was very evenhanded, not to say a bit ambiguous. It is very close to that of a number of contemporary ecologists, insofar as it takes account of many material factors (e.g., geography, climate, soil, etc.), as well as of economic and social ones. Man is seen as both above nature and as an integral part of it. They challenge not only vulgar materialism and absolute naturalism, but also the antienvironmentalist voluntarism that Stalinism installed for so many decades in the U.S.S.R. Their position moves back and forth within the space marked off by these two poles.

In spite of the deformations that Marxism has suffered and despite the present crisis it is experiencing, the influence of Marx and of Engels remains important in ecology, especially in political and social ecology. In my opinion, then, Marx and Engels have much to offer to ecology and to contemporary ecologists, just as do a number of other pioneers. But I do not believe that one should go so far as to say that ecology is actually a type of Marxism, or that Marx and Engels are major founders of ecology. They should take their place beside other forerunners, but they are not, strictly speaking, important founders of ecology.

NOTES

1. C. Fry, "Marxism versus Ecology," *Ecologist, 6*(9), November, 1976, 328–332; J. E. Chapel, Jr., "Marxism and Environmentalism," *Annals of the Association of American Geographers, 57*(1), March, 1967, 203–206; L. K. Caldwell, *Environment: A Challenge to Modern Society* (Garden City, N.Y.: Doubleday, 1971), 212: "Marxism, because of its acceptance of determinism as an explanation of human behavior, has not been very sympathetic to the ecological way of thinking about natural resources or the human environment."

2. M. Bookchin, "A Discussion on 'Listen Marxist,' " in *Post-Scarcity Anarchism* (Berkeley, Calif.: Ramparts Press, 1971), 223–246. See also his *Toward an Ecological Society* (Montreal: Black Rose, 1980), 193–210.

3. G. Biolat, *Marxisme et environment* (Paris: Editions Sociales, 1973); W. Mandel, "The Soviet Ecology Movement," *Science and Society, 36*(4), Winter, 1972, 285, 416; and A. Feeney, *Guardian,* January 25, 1984, 19.

4. H. L. Parsons, ed., *Marx and Engels on Ecology* (Westport, Conn.: Greenwood Press, 1977); C. Claude, "Marxisme/Ecologie," *Dialectiques, 31,* Winter, 1981, 116–122; K. D. Shifferd, "Karl Marx and the Environment," *Journal of Environmental Education, 3*(4), Summer, 1972, 39–42; G. Skirbekk, "Marxisme et ecologie," *Esprit, 42,* 1974, 643–652; reprinted as Chapter 6, this volume.

5. This kind of typology, with slight variations, is common in works on the ecological movement. See, among others, my book, *Mouvement écologiste, énergie et environment: Essais d'écosociologie* (Montreal: Editions coopératives Albert Saint-Martin, 1982), esp. 75–96 on the ecological movement in Quebec.

6. G. N. Syer, "Marx and Ecology," *Ecologist, 1*(16), October, 1971, 19–21; M. Clow, "Alienation from Nature, Marx, and Environmental Politics," *Alternatives, 10*(4), Summer, 1982, 36–40.

7. The book by A. Schmidt, *The Concept of Nature in Marx* (London: New Left Books, 1972), published in German in 1962, does not speak of ecology as such, but contains, like Parsons's book cited above, several important references.

8. K. Marx, *La différence de la philosophie de la nature chez Démocrite et Epicure* (Bordeaux: Ducros, 1970).

9. B. Commoner, *The Closing Circle: Nature, Man, and Technology* (New York: Alfred A. Knopf, 1971).

10. K. Marx, *Manuscripts de 1844* (Paris: Editions Sociales, 1972), 87.

11. Ibid., 101.

12. F. Engels, "Introduction to the Critique of Political Economy," cited in Parsons, *Marx and Engels on Ecology,* 172.

13. K. Marx and F. Engels, *Critique de Malthus* (Paris: Maspero, 1978), 58.

14. Ibid., 16.

15. Ibid., 16, 64.

16. Ibid., 65.

17. F. Engels, *Die Lage der arbeitenden Klassen in England* (Berlin: New Edition, 1952).

18. L. R. Villermé, *Tableau de l'état physique et moral les ouvriers employés dans les manufactures de coton, de laine et de soie* (Paris: Union Génerale d'Editions, 1971 [1840]). See also J. Y. Calvez, *La pensée de Karl Marx,* 7th ed. (Paris: Seuil, 1966), 26.

19. F. Engels, *La situation de la classe laborieuse en Angleterre* (Paris: Editions Sociales, 1961), 141.

20. K. Marx, "On the Jewish Question," in Karl Marx, *Early Writings* (New York: McGraw-Hill, 1964), 37. In fact, this work of Marx dates from 1843, and it was published in the *Annales Franco-Allemandes* in 1844.

21. K. Marx, *Fondements de la critique de l'économie politique,* 1857–1859 draft (Paris: Anthropos, 1968).

22. K. Marx, *Manuscripts de 1857–1858 (Grundrisse)* (Paris: Editions Sociales, 1980), Tome II, 96–97.

23. K. Marx and F. Engels, *Lettres sur les sciences de la nature* (Paris: Ed. Sociales, 1973), 19.

24. Ibid., 20.

25. Ibid., 21.

26. Ibid., 22.

27. Ibid., 35.

28. K. Marx, *Le Capital* (Paris: Garnier-Flammarion, 1969), 1:47, 141–142.

29. Ibid., 200.

30. Ibid., 305–306.

31. Ibid., 362–363.

32. Ibid., 660.

33. Marx and Engels, *Lettres,* 62.

34. Ibid., 70–71.

35. As several researchers have shown recently, Darwin's letter of November 13, 1880, in which he refuses to allow a book to be dedicated to him, was not addressed to Marx but to Aveling, his son-in-law, who wanted to dedicate to him his book *The Student's Darwin.* See, among others, M. A. Fay, "Marx and Darwin, Literary Detective Story," *Monthly Review, 31*(10), March, 1980, 40–57.

36. Marx and Engels, *Lettres,* 85. This passage reappears, almost literally, in F. Engels, *Dialectique de la nature* (Paris: Editions Sociales, 1968), 317.

37. Ibid., 89.

38. Ibid., 94–95.

39. In Marx and Engels, *Lettres,* 114.

40. K. Marx, *Le Capital* (Paris: Editions Sociales, 1976), 2:213.

41. Ibid., 3:111–113.

42. Ibid., 3:735.

43. Marx and Engels, *Critique de Malthus,* 135–136; see also K. Marx, *Théories sur la plus-value* (Paris: Editions Sociales, 1975), 2:129.

44. Marx and Engels, *Critique de Malthus,* 229; Marx, *Théories sur la plus-value,* 3:66.

45. F. Engels, *Anti-Dühring* (Paris: Editions Sociales, 1973 [1878]), 41.

46. Ibid., 100.

47. Ibid., 104.

48. After creating a conservative political movement, the German Monist League in 1904, of which he was later to be honorary president, Haeckel became even more conservative and anti-Marxist.

49. Marx and Engels, *Lettres,* 95.

50. Engels, *Anti-Dühring,* 335–336.

51. Engels, *Dialectique de la nature,* 42.

52. Ibid., 178–181.

53. Ibid., 181.

54. Ibid., 183.

55. Ibid., 223.

56. Ibid., 209, 228–229, 277–278.

57. Ibid., 315–317.

58. K. Marx, *L'idéologie allemande,* in *Oeuvres choisies* (Paris: Gallimard, 1963), 1:125.

Chapter 3

MARX AND RESOURCE SCARCITY

Michael Perelman

INTRODUCTION

Marx is widely believed to have failed to understand the importance of natural resources.[1] However, he unambiguously defined the labor process as the transformation of nature into objects of utility.[2] He denounced the German socialist movement for ignoring the role of nature. And he insisted that "labor is not the source of all wealth," arguing that "nature is just as much the source of use values."[3]

Even so, Marx had belittled the importance of scarcity in his youth. He predicted that future society would easily master natural resource production.[4] He even speculated that the rate of productivity in agriculture would rise faster than that in industry.[5]

Suddenly, however, the shortage of cotton during the U.S. Civil War shocked Marx into studying rent and scarcity. Still, he feared the political consequences of giving a prominent role to scarcity. For this reason, he subsumed his theory of scarcity into the concept of the "organic composition of capital," leading his readers to misunderstand both the importance of scarcity and the relevance of the composition of capital in his theory of capitalism. As a result, Marxists and others have failed to appreciate Marx's sophisticated theory of natural resources.

THE COTTON FAMINE

During 1862 the prices of important raw materials, especially cotton, reached their highest point since the Napoleonic Wars.[6] Within a three-week period in August and September 1862, cotton prices rose by 50 percent.[7]

Marx and Engels usually welcomed the prospect of a crisis in the cotton trade.[8] However, this one caused exceptional suffering. The majority of Lancashire cotton workers were thrown out of work,[9] threatening their health.[10] Even the *Times* fulminated against the heartlessness of the Cotton Lords.[11] (In 1866, as British industry turned to India for cotton, rice culture was restricted, resulting in the infamous famine of 1866, "which cost the lives of a million people in the district of Orissa alone."[12])

The year 1862 was also a time of "disheartening material suffering" for Marx personally.[13] The *New York Tribune* severed its tenuous relationship with Marx, and Engels was unable to supply him with much money (Engels's factory was working at only one-half capacity).

By August, Marx wished that he knew how to start a business, paraphrasing Faust: "Gray, dear friend, is all theory, and only business is green."[14] Before the end of the year, he informed his friend Kugelmann that his attempt at obtaining a job with a railway office failed because of his inadequate handwriting.[15] By the beginning of the next year, Marx's family lacked coal to heat the house and enough warm clothing to go outdoors.[16]

Engels's letters contained frequent complaints about his personal difficulties.[17] Prior to the Cotton Famine, Engels had maintained two separate residences: one for receiving his bourgeois friends, and one for himself and Mary Burns. With the onset of the crisis, he had to save on rent by living with Mary Burns full-time.[18]

THE RECOGNITION OF SCARCITY

In the midst of this crisis, Marx began his intensive research into rent theory and reversed his previously optimistic prognosis for capitalist agriculture[19] (presented in *Theories of Surplus Value*), only a few pages after an upbeat assessment of the prospects for the evolution of capitalist agriculture.[20]

Marx was aware of the importance of agriculture's role as an industrial supplier well before the Civil War.[21] For example, he had chided Ricardo for failing to recognize the importance of the industrial demand for agricultural production, especially cotton.[22]

Marx distinguished between what he referred to as "social" and "natural" productivity.[23] In the early stages of society, cheap raw-materials present themselves as a natural fertility of capital.[24] Primary products cost little effort because "nature . . . assists as a machine."[25] By contrast, in industry "nature builds no machines, no locomotives, railways, electric telegraphs, self-acting mules, etc."[26]

Still, Marx did not build on these insights until the Cotton Famine. Thereafter, Marx recognized the importance of resource availability to both capital and the accumulation process.[27] For example, he showed that the rate of surplus value was unaffected by the increasing difficulty in producing foodstuffs during the period 1799 to 1815 only because real wages fell, while labor was forced to work longer hours at a more intense pace.[28] Only Malthus, among all the classical economists, seems to have also noticed this relationship.[29] Elsewhere, Marx added cheap colonial imports and new technology to his list of causes offsetting the difficulty in producing food domestically.[30]

Marx suggested that although great industrial advances could be expected to continue, technical change in raw material production would not be sufficient to compensate for long-run natural resource exhaustion.[31] Eventually, an increasing proportion of social labor must be applied to the production of primary materials, despite the enormous advances in capitalist agricultural technology.[32] This proportion would increase further (he thought) because of improved technology that would diminish the portion of social labor needed in the production of machinery.[33]

THE SCOURGE OF MALTHUSIANISM

After 1862, Marx regarded scarcity as an important category, with substantial theoretical and political implications. He hoped that his critique of Malthusian rent theory would demonstrate "how the price of raw materials influences the rate of profit."[34] For Marx, the specter of Malthusianism had to be exorcised at all costs for political reasons. He confided to Engels that "the more I get into this crap, the more I am convinced that agricultural reform . . . will be the alpha and omega of the coming revolution. Otherwise Parson Malthus would be correct."[35]

Marx refused to admit the concept of scarcity directly in chapter 5 of *Capital.* He even failed to address the widespread hardship brought on by the Cotton Famine, except for dating specific events. His analysis of technological unemployment was excessively mechanical, almost undialectical. Because this chapter, in spite of its undeniable importance, was flawed, it detracted from the rest of his work. Perhaps sensing its shortcomings, Marx, whose book had previously evolved in a very Hegelian fashion,[36] abruptly turned to the section on primitive accumulation, which contains his analysis of the evolution of capitalist agriculture. Marx even avoided discussing scarcity directly in his most direct analysis of Malthus—his chapter on "The General Law of Capitalist Accumulation."

Marx, apparently, did not feel free to assert directly that shortages of raw materials were responsible for crises. His reasons seemed to be political in nature. In early 1863, Lassalle had just taken over the workers' movement in Germany.[37] Lassalle proposed that his "Iron Law of Wages" proved that workers could do nothing to improve their lot, except to appeal to the government to help them establish cooperatives.[38] To emphasize scarcity would play into the hands of the Malthusians, or, more precisely, the Lassallians.

For Marx, raw material shortages reflected the inability of capital to master the environment. He was confident that under socialism such problems could be overcome, but the specifics of that victory could not be given in detail. Otherwise, he believed that he would have to begin a series of endless debates about the specifics of the appropriate form of socialist organization, which would only divert energies from more important tasks.

Rather than risk getting bogged down in debates about the nature of scarcity, Marx adopted abstract categories that seemed to have little to do with natural resource scarcity, even when he was addressing the momentous impact of the Cotton Famine.

In an effort to demonstrate the importance of scarcity without giving support to Malthus, Marx adopted his category of constant capital as an indicator of scarcity.[39] In this respect, had he merely demonstrated the Keynesian lesson that the market could create unemployment and poverty, he would have been more successful. Had he shown that the combination of recurrent crises and scarcities can cause so much hardship for the working class; that the uprooting of traditional societies faster than they could be incorporated into the labor markets ensures poverty for a large portion of the working class; and that capital requires poverty and unemployment to maintain a tractable labor supply, his contribution to political economy would have been unquestioned.

Instead, Marx succeed in obscuring some of his most important insights and confusing generations of readers.

LONG-RUN ENVIRONMENTAL FAILURE AND THE NEED FOR SOCIALISM

In the wake of the Cotton Famine, Marx began to recognize the connection between the structure of market incentives and the broader fate of the environment. Scarcity exists (he argued) because of the inability of capitalism to utilize nature effectively rather than because of natural shortages.

Although agricultural science has taken great strides, Marx's general position was that, while land can be improved and (more generally) human potential is unlimited,[40] the social relations of capital stood in the way of agricultural progress. He noted that capitalists, seeking short-term gain and fearing long-term investments, avoid sinking money in improvements in forestry,[41] soil conservation,[42] and other durable investment projects. Furthermore, capitalists' single-minded pursuit of profit blinds them to the totality of natural processes. Engels prophetically warned: "Let us not, however, flatter ourselves overmuch on account of our human victories over nature. For each such victory nature takes its revenge on us."[43] And Marx made the sweeping observation:

> Capitalist production has not yet succeeded and never will succeed in mastering these (organic) processes in the same way as it has mastered purely mechanical or inorganic chemical processes. Raw materials such as skins, etc., and other animal products become dearer partly because the insipid law of rent increases the value of these products as civilizations advance. As far as coal and metal (wood) are concerned, they become more difficult as mines are exhausted.[44]

He concluded: "It is in the nature of capitalist production that it develops industry more rapidly than agriculture. This is not due to the nature of the land, but to the fact that, in order to be exploited really in accordance with its nature, land requires different social relations."[45]

Marx cited Frederick L. Olmsted's observations of cotton production in the southern United States as an example of the relationship between the social relations of production and agricultural progress.[46] For Marx:

> The moral history . . . concerning agriculture . . . is that the capitalist system works against a rational agriculture, or that a rational agriculture is incompatible with the capitalist system (although the latter promotes technical improvements in agriculture), and needs either the hand of the small farmer living by his own labour or the control of associated producers.[47]

Marx did not come by this conclusion casually. He took copious notes on Liebig, Johnston, and other agronomists who gave detailed accounts of the problems of soil exhaustion.[48] In his opinion, "The new agricultural chemistry in Germany, especially Liebig and Schonbein . . . are more important than all the economists put together."[49] Marx continued to display a keen interest in the analysis of soil fertility.[50] He was convinced that an agricultural crisis threatened the "apparently" solid English society."[51]

Marx's agricultural research eventually led him to the verdict that capitalist agriculture "leaves deserts behind it."[52] His section on "Modern Industry and Agriculture" in the first volume of *Capital* reads like some of the most insightful literature from the modern environmental movement:

> Capitalist production collects the population together in great centres, and causes the urban population to achieve an ever-growing preponderance. This has two results. On the one hand it concentrates the historical motive power of society; on the other hand, it disturbs the metabolic interaction between man and the earth, i.e., it prevents the return to the soil of its constituent elements by man in the form of food and clothing; hence it hinders the operation of the eternal conditions for the lasting fertility of the soil. Thus it destroys at the same time the physical health of the urban worker, and the intellectual life of the rural worker. But by destroying the circumstances surrounding that metabolism, which originated in a merely natural and spontaneous fashion, it compels its systematic restoration as a regulative law of social production, and in a form adequate to the full development of the human race. . . . In modern agriculture, as in urban industry, the increase in the productivity and mobility of labour power is purchased at the cost of laying waste and debilitating labour power itself. Moreover, all progress in capitalist agriculture is a progress towards ruining the more long-lasting sources of that fertility. The more a country proceeds from large-scale industry as the background of its development, as in the case of the United States, the more rapid is this process of destruction. Capitalist production, therefore, only develops the techniques and the degree of combining of the social process of production by simultaneously undermining the original sources of all wealth—the soil and the worker.[53]

Marx even went so far as to speculate that the destruction of the land by the mindless profit-seeking bourgeoisie represented "another hidden socialist tendency."[54] Sooner or later, he predicted, capital will discover that nationalization of the land would be the only course capable of assuring an adequate supply of agricultural produce.[55] Quite a performance for a nineteenth–century writer who is supposed to have insisted that nature is unimportant!

MARX, MALTHUS, AND THE ORGANIC COMPOSITION OF CAPITAL

Marx was certain that raw materials were the most important component of constant capital, at least on a flow basis.[56] He insisted that raw materials, especially cotton, were "the most important element in all branches" other than wages.[57] This importance was likely to grow. In addressing the increasing difficulties in producing enough raw material for industrial production, more often than not he brought up the example of cotton. Significantly, in Marx's description of constant capital, "raw material, auxiliary materials, machinery, etc.," raw materials were given first place.[58]

Although Marx had earlier alluded to the possibility that rising cotton prices began to represent an increase in the organic composition of capital,[59] the Cotton Famine made that possibility a reality. From that point on, Marx devoted a good deal of attention to the organic composition of capital. More often than not, he used cotton rather than machinery as the primary example of the rising organic composition of capital.[60] In a section entitled "On the Influence of a Change in the Value of Constant Capital Exerts on Surplus, Profit and Wages," he concluded that "this analysis shows the importance of the cheapness or dearness of raw materials for the industry which works them up (not to speak of the relative cheapening of machinery)."[61]

In a footnote to this citation, he suggested the nature of his mental association between the discussion of the analysis of cotton prices and the analysis of the organic, value, and technical compositions of capital: "By *relative* cheapening of machinery, I mean that the absolute value of the amount of machinery employed increases, but that it does not increase in the same proportion as the mass and efficiency of machinery."[62]

The "organic" in the expression "organic composition of capital," suggests a biological dimension. In fact, the earliest use of that term occurred in the *Theories of Surplus Value.* In January 1863, in the midst of the Cotton Famine, Marx used this term in an outline of his later treatment of the falling rate of profit.[63] There he wrote:

> 1. Different organic compositions of capitals, partly conditioned by the differences between variable and constant capital in so far as this arises from the *stage of production*–the absolute *quantitative* relations between machinery and raw materials on the one hand, and the quantity of labour which sets them in motion. These differences relate to the labour process.
> 2. Differences in the relative value of the parts of different capitals which do not arise from their organic composition. These arise from the difference of value particularly of the raw materials, even assuming that the raw materials absorb an equal quantity of labour in two different spheres.[64]

Note several points about this discussion. First, raw materials and machinery are lumped together. Second, the reference to the "relative value of the parts . . . which do not rise from their organic composition" suggests that Marx had in mind two different causes for the value of raw materials to change. On the one hand, value changes because of the changing labor requirements. On the other hand, relative values can fluctuate because of other forces. Coming on the heels of the enormous upheavals stemming from speculation in the cotton market, Marx may well have been thinking about the price instability of raw materials.

Does the frequency with which Marx associated the rising organic composition of capital with cotton suggest that his theory of the rising organic composition of capital is an obscure but convenient method of introducing into his analysis an important phenomenon, one usually considered to be Malthusian?

In a later section, dealing with the subject of "Compound Interest: Fall in the Rate of Profit Based on This," Marx referred to the "organic ratio between constant and variable capital. In other words, the increase in the capital in relation to labour is here identical with the increase of constant capital as compared with variable capital and, in general, with the amount of living labour employed."[65]

The context of this reference to the "organic ratio" is especially revealing. Marx began the section by addressing the possibility that technical change might so effectively cheapen the cost of living that the rate of profit might rise. Marx responded: "The value of labour-power does not fall in the same degree as the productivity of labour or of capital increases."[66] Why not? Marx answered: "It is in the nature of capitalist production that it develops industry more rapidly than agriculture."[67]

This disparity between agriculture and industry was reflected in the turnover rates in their respective capital stocks. Concerning this matter, Marx wrote:

> The period necessary to get the product ready for the market . . . is based on the existing material conditions of production specific for the various investments of capital. In agriculture they assume more of the character of the natural conditions of production, in manufacture and the greater part of they mining industry they vary with the social development of the process of production itself.[68]

Here it is important to recall the importance of the turnover rate of capital as a determinant of the rate of profit.

Marx continued, in his published works, to associate the organic composition of capital with the production of raw materials, especially cotton. He first mentioned the organic composition of capital (*sans* organic) immediately before writing:

> As to raw materials, there can be no doubt that the rapid advance of cotton spinning not only promoted as if in a hot house the growing of

cotton in the United States, and with it the African slave-trade, but also made slave-breeding the chief business of the so-called border slave states.[69]

Marx first introduced the complete term "organic composition of capital" to the public in chapter 25 of the first volume of *Capital,* the same chapter in which he contrasted his law of the demand for labor with Malthus's theory of population. The first sentence reads: "In this chapter we shall consider the influence of the growth of capital on the fate of the working class."[70]

Two paragraphs later, he defined the organic composition of capital. Immediately thereafter, he stated: "If we assume that . . . the composition of capital remains constant, then the demand for labour . . . [will] clearly increase in the same proportion and at the same rate as capital."[71]

Next Marx turned to a critique of classical political economy, "Adam Smith, Ricardo, etc."[72] Malthus was conspicuously absent. The central idea of this discussion was that the poor were necessary to maintain the rich. In other words, populationism was not a sufficient explanation of poverty.

A few pages later, in a section entitled "The Progressive Production of a Relative Surplus Population or Industrial Reserve Army," Marx alluded to Malthus less obliquely. He had been pressing his theory that the increase in constant capital reduces the demand for variable capital: "The working population therefore produces both the accumulation of capital and the means by which it is itself made relatively superfluous. . . . *This is a law of population* peculiar to the capitalist mode of production."[73] Marx then continued: "An *abstract law of population exists only for plants and animals* and even then only in the absence of any historical intervention by man."[74] Malthus is mentioned only twice in this chapter. On one occasion, Marx wrote, "Even Malthus recognizes that a surplus population is a necessity of modern industry."[75] On the other occasion, the reference to Malthus is contained in a long footnote that began: "If the reader thinks at this point of Malthus . . . , I would remind him that this work in its first form is nothing more than a schoolboyish plagiarism."[76]

Here we find Malthus largely absent from the discussion of Malthusianism. Instead, we find an introduction to the organic composition of capital.

THE FALLING RATE OF PROFIT

I have been suggesting that Marx's unwillingness to lend political credence to Malthusianism led him to obscure his analysis of scarcity. If this hypothesis is correct, it does provide a solution to an important riddle.

For Marx, the law of the falling rate of profit was "in every respect the most important law of modern political economy, and the most essential for understanding the most difficult relations."[77] I am convinced that his quest for a theory of the falling rate of profit became entangled with his analysis of scarcity. By treating scarcity via profit rates, he could avoid the grave political risks of debating Malthusianism. This tactic reinforced his quest for an automatic law of the falling rate of profit.

Many modern commentators are struck by the relative lack of sophistication in Marx's rather formalistic two-sector theory of the tendency of the rate of profit to fall. This anomaly is especially obvious when one compares his theory of the falling rate of profit with the mathematical virtuosity that he displayed in his two sector models in the second volume of *Capital.*

Siegel has pointed out that Marx's section on the falling rate of profit contained numerous uncharacteristic slips, which suggest that Marx's own doubts about this part of his theory persisted despite his strong commitment to discover a law that necessitated a fall in the rate of profit.[78] Engels's skepticism about this part of Marx's theory led him to edit this section in a way that minimized the deterministic aspects of this part of Marx's analysis.[79] Moreover, Engels himself, even though he was the chief interpreter of Marxist theory for twelve years after Marx's death, never wrote anything about the falling rate of profit.[80]

Scarcity may be more significant in Marx's theory than has been previously suspected. I propose that Marx had used the organic composition of capital as a code for scarcity and that, in the back of Marx's mind, scarcity was responsible for the falling rate of profit.

If my hypothesis is correct, Marx's tactic impoverished Marxist theory. The rehabilitation of Marx's notion of scarcity opens Marxist theory up to a rich line of analysis that can help to put Marxist analysis at the forefront of the inevitable future debates about the political economy of natural resource utilization.

REVALUATION AND SHORT-RUN CRISES

The cotton industry, which loomed so large in Marx's analysis, bore out his conviction that raw materials production would have difficulty keeping pace with demand. For example, in the brief period between 1830 and 1837, as industry had outstripped agricultural production, cotton prices doubled, only to fall again.[81]

This instability of raw materials prices reflected another dimension of the role of natural resource shortages within capitalism. In addition to the difficulties created by the long-run tendency of scarcity, erratic short-term raw materials prices occur because the short-run supply elasticity of agricultural produce makes trade in such items particularly vulnerable to speculative pressures, leading to "sudden expansion soon followed by collapse."[82] In the 1862 Cotton Famine, Marx first associated these short-term crises and raw materials shortages. At this time, when *Capital* was being written, cotton again became scarce. This pattern moved Marx to observe:

> It is in the nature of things that vegetable and animal substances whose growth and production are subject to certain organic laws and bound up with definite natural time periods, cannot be augmented in the same degree as . . . other fixed capital . . . , whose reproduction can, provided the natural conditions do not change, be rapidly accomplished in an industrialised country. It is therefore quite possible, and under a developed system of capitalist production even inevitable, that the production and increase of the portion of constant capital consisting of fixed capital, machinery, etc. [measured in physical terms], should considerably outstrip the portion consisting of organic raw materials so that demand for the latter rises more rapidly than supply. . . .
>
> The greater the development of capitalist production . . . , so much more frequent the relative underproduction of vegetable and animal raw materials, and so much more pronounced the previously described rise of their prices and the attendant reaction. And so much more frequent are the convulsions caused as they are by the violent price fluctuations of one of the main elements in the process of reproduction.[83]

Marx observed:

> A *crisis* can rise . . . through *changes in the value of the* elements of productive capital, particularly of *raw materials,* for example when there is a decrease in the quantity of cotton harvested. Its *value* will thus arise. . . . More must be expended on raw materials, less remains for labor.[84]

Even though the value of raw materials might be less than the value of the stock of fixed capital, a rise in raw material prices can make itself felt as a decline in the profit rate, with serious consequences for accumulation:

> If the price of raw materials rises, it may be impossible to make it good fully out of the price of commodities after wages are deducted. Violent price fluctuations, therefore, cause interruptions, great collisions, even catastrophes, in the process of reproduction. It is especially agricultural produce proper, i.e., raw materials taken from organic nature, which—

leaving aside the credit system for the present—is subject to such fluctuations in values in consequence of changing yield, etc.[85]

An actual improvement of raw materials satisfying not only the desired quantity, but also the quality desired, such as cotton from India of American quality, would require a prolonged, regularly growing and steady demand (regardless of the economic conditions under which the Indian producer labours in his country). As it is, the sphere of production is, by fits, first suddenly enlarged, and then violently curtailed. All this, and this spirit of capitalist production in general, may be very well studied in the cotton shortage of 1861–65, further characterised as it was by the fact that a raw material, one of the principal elements of reproduction was for a time unavailable.[86]

Increasing natural resource costs are frequently cited as proof of the operation of the law of diminishing returns, but Marx interpreted the same phenomenon rather differently; he saw it as evidence of a barrier posed by capitalist social relations. For example, Marx insisted that the booms and busts that are endemic to capitalism are incompatible with a rational agricultural system:

The closer we approach our own time in the history of production, the more regularly do we find, especially in the essential lines of industry, the ever-recurring alteration between relative appreciation and the subsequent resulting depreciation of raw materials obtained from organic nature.[87]

This secular tendency is paralleled by cyclical difficulties in raw materials production. As the economy expands, the rapid growth in demand for raw materials, generated by capitalist growth, will not be matched by a proportionate increase in the production of raw materials.[88]

References to the changing values of raw materials recur in Marx's later work. He placed a similar thought in a section entitled "Observations on the Influence of the Change in the Value of the Means of Subsistence and of Raw Material (Hence also the Value of Machinery) on the Organic Composition of Capital."[89] In the next section, he returned to the subject of the organic composition of capital to explain how it affected the pricing of agricultural produce. In addition, chapter 6 of the third volume of *Capital* largely concerned the effect of raw material price fluctuations on the value of capital.

As prices collapse, the organic composition of capital recedes and those who have unrealized appreciation on their stocks of raw materials see their potential gains disappear. Unlike long-term crises, the short-term crises can be self-correcting—if the system can withstand the shock.

CONCLUSION

Unemployment or poverty cannot be reduced to natural laws; rather, they reveal fundamental flaws in capitalist society. Consequently, observations of raw material shortages, soil erosion, rising food prices, or apparent overpopulation elicited a consistent political and methodological response from Marx. What appears as a "Malthusian problem" was, in reality, a reflection of a contradiction within capitalist society.

Such problems concerning agriculture and the appropriation of natural resources first seem to have drawn Marx's attention to social questions.[90] Marx saw that such problems opened up enormous opportunities for social and political change. For example, a rise in raw material prices was one of the preconditions of Louis Napoleon's coup in France.[91]

In this vein, Marx taught that the "ever growing wants of the people on one side" and "the ever increasing price of agricultural produce on the other" offered an excellent opportunity to organize for the nationalization of the land.[92]

Lenin stood firmly in this tradition. He observed that "the more capital is developed, the more strongly the shortage of raw materials is felt, the more intense the competition and hunt for sources of raw materials throughout the world, the more desperate the struggle for the acquisition of colonies."[93] He added that "to try to belittle the importance of facts of this kind by arguing that . . . the supply of raw materials "could be" increased enormously by "simply" improving the conditions of agriculture" would be to repeat the mistakes of bourgeois reformists such as Kautsky.[94]

To carry on with Marx's project is no mean task. It is made no easier by the methodological tools with which we have saddled ourselves. Terms such as "scarcity," "shortage," or "depletion" conjure up images of technical needs. They suggest that if only more oil or better methods for handling resources were available, then economic problems would disappear. This perspective leads in circles. Each new technique is followed by new problems and the need for still newer techniques.

Marx attempted to forge a new set of categories to analyze these "Malthusian problems" to the betterment of human society.[95] In place of overpopulation, he referred to the reserve army of the unemployed. Instead of allowing us to become bogged down in an ahistorical concept of resource scarcity, he tried to grasp the social content of each situation. Marx's distinction between "historically developed" and "naturally conditioned productive forces" illustrates the manner in which "natural" and organizational phenomena are bound together.[96] The frequency with which he used examples of natural resource scarcity, when explaining the organic composition of capital, suggests another methodological alternative to the concept of scarcity he may have had in mind. His work was never completed. It is left to us to carry on.

Marx's work on the social relations of natural resource scarcity can be of great use in approaching the contemporary problems of natural resource depletion and contamination which plague our society today. Living in an age in which cost–benefit analyses coldly calculate "appropriate" levels of destruction of our heritage of natural resources, Marx's analysis points the way toward acting in a way to preserve our birthright and create a humane society.

ACKNOWLEDGMENT

The author wishes to acknowledge the patient assistance of James O'Connor in preparing this chapter.

NOTES

1. Paul A. Samuelson, "Wages and Interest—A Modern Dissection of Marxian Economic Models," *American Economic Review, 67*(6), December, 1967, 894.

2. Karl Marx, *Capital* (New York: Vintage, 1977), 1:chap. 7.

3. Karl Marx and Frederich Engels, "Critique of the Gotha Program," in *Selected Works in Three Volumes* (Moscow: Progress Publishers, 1973), 3:13; see also Marx, *The Economic and Philosophical Manuscripts of 1844* (New York: International Publishers, 1964), 109.

4. Karl Marx, *Theories of Surplus Value*, 3 vols. (Moscow: Progress Publishers, 1963–1971), 1:219.

5. Ibid., 2:109-112; see also Marx, *Capital* (New York: International Publishers, 1967), 3:760; and Marx to Engels, August 2, 1862, in Karl Marx and Frederich Engels, *Selected Correspondence* (Moscow: Progress Publishers, 1975), 123.

6. D. A. Farnie, *The English Cotton Industry and the World Market, 1815–1896* (Oxford: Clarendon Press, 1979), 162.

7. Ibid., 145.

8. Karl Marx and Frederich Engels, *Collected Works,* Vol. 40: *Letters: January, 1856–December, 1859* (New York: International Publishers, 1983), 221; Engels to Marx, December 9, 11, and 17, 1857, and October 17, 1858, *Letters,* 220–223, 343–345).

9. Marx, *Capital,* 1:720; Marx, "Garibaldi Meetings—Distressed Condition of Cotton Workers," *Die Presse, 273,* September, 1862 (reprinted in *Collected Works,* Vol. 19: *Marx and Engels, 1861–1864* [New York: International Publishers, 1984], 245–247).

10. Engels to Marx, November 5, 1862, in Marx and Engels, *Selected Correspondence,* 295; W. O. Henderson, *The Life of Friedrich Engels* (London: Frank Cass, 1976), chap. 5; Farnie, *English Cotton Industry,* 157.

11. Marx, "Garibaldi Meetings."

12. Marx, *Capital*, 2:141.

13. M. Rubel and M. Manale, *Marx without Myth* (New York: Harper and Row, 1975), 174.

14. Marx to Engels, August 20, 1862, in Marx and Engels, *Selected Correspondence*, 280.

15. Marx to Kugelmann, December 28, 1862, *Selected Correspondence*, 640.

16. Marx to Engels, January 24, 1863, in *Selected Correspondence*, 314–316.

17. Engels to Marx, July 3, 1862, in Marx and Engels, *Collected Works*, Vol. 41: *Letters: January, 1860–December, 1864* (New York: International Publishers, 1985), 382, 413, 414, 418.

18. Engels to Marx, February 28, 1862, in Marx and Engels *Collected Works*, 30:215.

19. Marx to Engels, August 2, 1862, in Marx and Engels, *Selected Correspondence* (1975), 120–123.

20. Marx, *Theories of Surplus Value*, 2:109–115.

21. Karl Marx, *Grundrisse* (New York: Vintage Books, 1974), 728, 771ff.

22. Ibid., 640.

23. Marx, *Capital*, 3:766.

24. Ibid., 106, 651.

25. Marx, *Theories of Surplus Value*, 2:109; Marx, *Capital*, 1:284, 744; 3:360–361, 643, 745, 847–848.

26. Marx, *Grundrisse*, 706, 715.

27. Marx, *Theories of Surplus Value*, 2:516; Marx, *Capital*, 1:579.

28. Ibid., 666; Marx, *Theories of Surplus Value*, 3:408.

29. Marx, *Capital*, 1:666fn.

30. Marx, *Theories of Surplus Value*, 2:460.

31. Ibid., 766; Michael Perelman, "Natural Resources and Agriculture: Karl Marx's Economic Model," *American Journal of Agricultural Economics*, 57(4), November, 1975, 701–704.

32. Marx, *Capital*, 3:745; 1:751.

33. Marx, *Capital*, 3:109.

34. Marx to Engels, April 30, 1868, in Marx and Engels, *Selected Correspondence* (New York: International Publishers, 1942), 242.

35. Marx to Engels, August 14, 1851, in Marx and Engels, *Selected Correspondence*, 314.

36. Michael Perelman, *Marx's Crisis Theory: Scarcity, Labor, and Finance* (Westport, Conn.: Praeger, 1987), chap. 4.

37. Hal Draper, *Karl Marx's Theory of Revolution*, Vol. 4: *Critique of Other Socialisms* (New York: Monthly Review Press, 1990), 261.

38. Ibid., 52.

39. Perelman, *Marx's Crisis Theory*, chap. 5.

40. Marx, *Theories of Surplus Value*, 2:144–45, 595.

41. Marx, *Capital*, 2:235; 1:892–893.

42. Marx, *Capital*, 3:617; 1:376.

43. Friedrich Engels, "The Part Played by Labour in the Transition from Ape

to Man," in Marx and Engels, *Selected Works in Three Volumes* (Moscow: Progress Publishers, 1970), 1:74–75.

44. Marx, *Theories of Surplus Value*, 3:368.

45. Marx, Ibid., 300–301.

46. Cited in Marx, *Capital*, 1:304fn.

47. Marx, *Capital*, 3:121.

48. Marx, *Grundrisse*, 754fn.

49. Marx to Engels, February 13, 1866, in Marx and Engels, *Selected Correspondence* (1942), 204–205.

50. Marx to Danielson, February 19, 1881, in *Selected Correspondence* (1942), 384.

51. Marx to Danielson, April 10, 1879, in Marx and Engels, *Selected Correspondence* (1975), 298.

52. Marx to Engels, March 25, 1868, in Marx and Engels, *Selected Correspondence* (1942), 237.

53. Marx, *Capital*, 1:636–638; Marx, *Capital*, 3:301.

54. Marx to Engels, March 25, 1868, in Marx and Engels, *Selected Correspondence* (1942), 237.

55. K. Marx, "Nationalization of Land," in Marx and Engels, *Selected Works in Three Volumes* 2:288–290.

56. Marx, *Capital*, 3:106.

57. Ibid., 117; see also 106.

58. Ibid., 120.

59. Marx, *Theories of Surplus Value*, 1:195.

60. Ibid., 3:chap. 23, and 217–221.

61. Ibid., 3:221.

62. Ibid.

63. Ibid., 1:415–416.

64. Ibid.

65. Ibid., 3:311.

66. Ibid., 300.

67. Ibid., 300–301.

68. Marx, *Capital*, 2:316.

69. Marx, *Capital*, 1:571.

70. Ibid., 762.

71. Ibid., 763.

72. Ibid., 764.

73. Ibid., 784.

74. Ibid., 784; emphasis added.

75. Ibid., 787.

76. Ibid., 766.

77. Marx, *Grundrisse*, 748.

78. Jerold Siegel, *Marx's Fate: The Shape of Life* (Princeton, N.J.: Princeton University Press, 1978), chap. 11.

79. Ibid.

80. John E. King, "Friedrich Engels as Economist: The Last Twelve Years, 1883–1895" (unpublished manuscript, 1985).

81. Peter Temin, *The Jacksonian Economy* (New York: W. W. Norton, 1969), 92.

82. Marx, *Capital*, 2:316.

83. Ibid., 3:118–119.

84. Marx, *Theories of Surplus Value*, 2:515; see also 517, 533.

85. Marx, *Capital*, 3:117.

86. Ibid., 120–121.

87. Ibid., 121.

88. Marx, *Theories of Surplus Value*, 2:533; Marx, *Capital*, 1:579.

89. Ibid., 275ff.

90. Karl Marx, *A Contribution to the Critique of Political Economy* (New York: International Publishers, 1970), 19–20; Engels to Richard Fischer, April 15, 1895, in Marx and Engels, *Capital*, 39:466–467.

91. Marx, 1969, 287.

92. Marx, "Nationalization of Land," 289.

93. V. I. Lenin, "Imperialism, the Highest Stage of Capitalism," in Lenin, *Collected Works* (Moscow: Progress Publishers, 1964), 22:260.

94. Ibid., 261.

95. David Harvey, "Population, Resources, and the Ideology of Science," *Economic Geography, 50*(3), July, 1974, reprinted in Richard Peet, ed., *Radical Geography* (Chicago: Maaroufa, 1977).

96. Marx, *Capital*, 1:651.

Chapter 4

GREENING PROMETHEUS
Marxism and Ecology

KATE SOPER

For the first time in history, ecology is moving to the center stage of politics. Its arrival there is long overdue and may have come too late. But it seems unreasonable now to doubt that so long as a forum for politics remains, it will be deeply involved with matters of ecology. Indeed, it is now clear that the fate of the planet and its various species, including our own, hangs on the nature and depth of the "green revolution" we manage to effect.

It is this center-staging of ecology that has inclined many to herald the green movement as the most important radical force for social change since the development of socialism, and particularly Marxist socialism, in the nineteenth century.[1] The comparison is telling insofar as it captures the here-to-stay quality of ecological pressure: just as nothing was ever the same again after the *Communist Manifesto,* so we can safely assume that no politics in the future will be able to ignore the Greens. It also brings out the contrast between the attention now being paid to ecological arguments and hedonist projections, and the almost total neglect in the past of green alarms and visions of utopia.[2]

However, the counterposing of socialism to ecology that is implied in

81

this approach to the coming of the Greens is very misleading. For just as socialism can only hope to remain a radical and benign pressure for social change by assuming an ecological dimension, so the ecological concern will remain largely ineffective (and certainly incapable of reversing the current trends in the manner required) if it is not associated in a very integral way with many traditional socialist demands, such as assaulting the global stranglehold of multinational capital.

Obviously, there are now many avowed "ecosocialists" who recognize this truth and are committed to the "red–green" synthesis. Some would even insist that the "socialism" in question here should be given a strongly Marxian interpretation (which does not at all mean associating it with the practices of "actually existing socialism"). Others, I think, would prefer simply to make clear that the "socialism" they are advocating is neither the laborism or social democratic reformism associated with the European socialist parties, nor Soviet-style communism. But there are many "Greens," both in and out of the green parties, who are very suspicious of the use of any socialist, let alone Marxist, vocabulary; and some, too, I believe, among the "ecosocialists" who remain very uncomfortable with an explicitly Marxian form of argument. It is this sense, fairly pervasive in green circles, that even if socialism and ecology might recognize a kinship, Marxism and ecology are much harder to splice and may even belong to hostile tribes, that sets the agenda for this chapter. I am not attempting here to deal with the range of issues raised under the broader "ecosocialist" concept, but am instead proposing to confine myself to the narrower task of considering the relations between ecological and specifically Marxist arguments. In what follows, then, I shall discuss the reasons behind the resistance among green thinkers to Marxism, consider how far they are justified, and put forward some theories of Marxist argument that need to be weighed in the balance before any final judgment is delivered on Marx's ecological credentials.

WHY THE MARX ALLERGY?

There are three main reasons why Marxism has been thought to be no friend of ecology. Of these, the most important is probably not primarily intellectual but political. It derives from the association of Marxism with the societies of "actual existing socialism" and their communist party support. It derives, that is, from the link between Marxism and the politics (until very lately) of Stalinism and neo-Stalinism, and the associated economic practice that has brought some of the worst pollution and environmental degradation in the industrial world to the countries of Central and Eastern Europe.[3] Large parts of Poland and Czechoslovakia are now designated ecological disaster zones, and the situation is not much better in the other former Soviet-bloc countries, including some parts of Russia itself. The problem, of course, does not stem in some abstract fashion from the imposition of socialist methods of economic organization, but instead from their use in a "catchup-and-overtake" pursuit of capitalist styles and levels of consumption.[4] Nevertheless, it was this particular use of the powers of socialist planning, together with its totalitarian political integument, that was foisted upon the peoples of Central and Eastern Europe in the name of "Marxism-Leninism," and the association between the practice and its supposed "theory" is one that is understandable, even if Marx and Engels would have been amazed and dismayed by the "socialism" to which their names have been annexed.

The second main reason for ecological suspicion of Marxism is more theoretical. It relates to the argument of historical materialism itself, which seems to accord priority to the "development of the productive forces" as a criterion and goal of social progress, which makes appeals to an era of communist "abundance," and which attaches importance to the advanced technological infrastructure that capitalism will bequeath to the socialist postrevolutionary forces of renewal. The "productivist" dynamics of this portrayal of human development, the associated faith in the virtues of technical growth, and the implied anthropocentric and instrumental attitudes to the rest of nature—all this is more or less diametrically opposed, it seems, to the green emphases on sustainability, on *re*productive rather than growth economies, on *de*industrialization, on sober consumption, and on a more humble and holistic approach to world ecological balance.

Of course, as we shall see in what follows, the "productivist" and "technocratic" Marx is not the only Marx. Moreover, even those texts (e.g., the 1859 "Preface" to the "Critique of Political Economy") that lend themselves most directly to a "technological determinist" reading of historical materialism[5] demand a more qualified and skeptical appraisal as guides to the "true" Marxist theory when placed in the context of Marx's work as a whole. Nonetheless, although it may be a caricature in certain respects, the "green" perception of historical materialism as antiecological is not entirely without justification; there are even some self-styled Marxists who would defend the less-green aspects of Marxism as distinguishing its own point of view from the Arcadian nostalgias of the socialist utopians. By the same token, Greens find, unsurprisingly, a deep vein of incompatibility between their own and the Marxist projections of progress and its means of achievement. Some of these points have been brought together in Rudolf Bahro's argument:

> Our customary idea of the transition to socialism is the abolition of the capitalist order within the basic conditions European civilization has created in the field of technique and technology—and not in Europe

alone. Even in this century, a thinker as profound as Antonio Gramsci was still able to view technique, industrialization, Americanism, the Ford system in its existing form as by and large an inescapable necessity, and thus depict socialism as the genuine executor of human adaptation to modern machinery and technology. Marxists have so far rarely considered that humanity has not only to transform its relations of production, but must also fundamentally transform the entire character of its mode of production, i.e. the productive forces, the so-called technostructure. It must not see its perspective as bound up with any historically transmitted form of the development of needs and their satisfaction, or of the world of products designed for that purpose. The commodity world that we find around us is not in its present form a necessary condition of human existence. It does not have to look the way it does in order for human beings to develop both intellectually and emotionally as far as we would like.[6]

Third, many Greens would challenge the adequacy and relevance today of the Marxist class-based analysis and revolutionary strategy.[7] How, it is asked, when we are dealing with such universal, transclass problems as those of resource attrition, the depletion of the ozone layer, Arctic pollution, the "greenhouse effect," and so forth, can it be thought appropriate to approach them purely in class terms or to suggest that we must see through the class struggle to its goal of proletarian emancipation before we can hope for any satisfactory resolution of the ecological crisis? In any case, if emancipation is conceived—as it often has been in communist party circles—in terms of increased productivity and expanded consumption, it must exacerbate rather than counter resource exhaustion and pollution. Indeed, even when it comes to the management of the ecological crisis within any nation-state, and the adjustment of its social inequalities, the Marxist recipe of "working-class victory" is regarded as increasingly unconvincing given the fragmentation of the traditional proletarian class, the economic advancement of many of its members relative to those occupying traditional petit-bourgeois and middle-class jobs, and the political impotence of all the most economically and socially deprived sectors of contemporary capitalist society. Industrial workers, moreover, driven by the logic of capital into putting job security before every other consideration, are by no means the first constituency to which one looks for protection of the environment—a point brought home rather forcefully by the posting of an unofficial blacklist at the Trawsfynydd power station by which union members were advised which local shops had put up posters for an antinuclear meeting.[8] Not all Greens would agree that these developments have rendered any form of class politics obsolete. Many would accept, in fact, the idea that ours is still very much a class society, and that the capitalists are, as ever, firmly in control and having the best of everything.

But even the more ecosocialist vein within the green movement would insist that the orthodoxy of "class struggle" needs considerable rethinking if it is to be anything more than a ritual response to the grotesque exploitations—both human and ecological—of our times.

OF ALIENATION AND FETISHISM

I shall in a moment suggest some reasons for qualifying the picture of Marxism as inherently unfriendly to green arguments. Nonetheless, I think it would be doing a disservice to Marx and Marxism to attempt to claim that there is really no target at all for the green attack. The Marxist corpus is a large one, and it is always possible to bend the stick of interpretation, dispute the meanings placed on various texts, and so on, and come up with a somewhat different picture. But that is not the point. The point is rather that there are without doubt significant elements of the Marxist argument that either directly conflict with much green thinking or are too undereexlaborated for us to claim definitively that they do not. Inversely, however, I would argue that a disservice is done to the green cause by all those who refuse to recognize these ambiguities in Marxism or to admit that there are any aspects at all of its outlook that are more consonant with their own position. For there certainly are themes in Marx that are not only congruent with current ecological critique, but that powerfully reinforce it. They are also of some importance in what they imply for the economic and political strategies essential to its progress.

In the first place, there is the definite green slant of the arguments regarding the concept of "alienation" (and the related notions of "reification" and "fetishism"). When green critics make use of the term "alienation" to describe the bleak environments and impersonal relations resulting from industrialization they may acknowledge a debt to Marxist theory. But the deeper pertinence of the theory of alienation is seldom recognized. For what Marx meant by "alienation" (and his usage of the term is here formally consistent with Hegel's)[9] is a process whereby a "subject" conceives "itself as subservient to an external and supposedly quite independent objectivity" of which "it" is "itself," in reality, the creative source. If we think of the "subject" here as the contemporary market society, then there is no doubt of the extent to which "it" assumes "itself" to be held in thrall to those very same economic determinations (financial investments, interest rates, inflationary tendencies, etc.) that "it" is "itself"—through the cumulative acts of its various individual members—responsible for setting in motion. When green critics, for example, challenge the rhetoric that speaks of the "critical condition" of the pound as if it were a patient in a hospital bed (now rallying and able to take some nourishment from the decline of the dollar, now falling back, etc.); when they denounce the perversion of a society that needs to export more Cindy dolls before it can afford the luxury of child kidney dialysis; when they point out how far economic forces are in the saddle and riding us—what else are they doing if not exposing what Marx referred to as capitalist "alienation?"[10]

Let us not forget, either, that the attack on alienation goes together with a most passionate denunciation of the fetishism of money and of the subordination through the "cash-nexus" of all other values to those of commercial profit. "Money," thunders Marx in the *1844 Manuscripts* is the "general distorting of all individualities and conversion of them into their opposites"; it "transforms fidelity into infidelity, love into hate, virtue into vice, vice into virtue, servant into master, master into servant, idiocy into intelligence and intelligence into idiocy. . . . It is the world upside-down—the confounding of all natural and human qualities."[11]

These and other similar polemics against the inverted values of bourgeois society may not be green in the strict sense of relating explicitly to the destruction of the natural environment. But they are certainly fired with the same sense of revolt against the philistine and mercenary tendencies of the market society that moves so much of the green critique at the present time.

There is also, at the very least, a family resemblance between much contemporary green argument and Marx's more theoretical discussion around the notion of fetishism in the chapter on "commodity fetishism" at the beginning of *Capital.* Jeremy Seabrook, for example, has recently drawn attention to the "magic of the markets" through which the "blood and filth and pain" that attend the creation of the products of Western affluence are miraculously washed away by the money that can afford to buy them:

> One of the most extraordinary by-products of the information-rich society is the creation of a kind of un-knowing, even ignorance, that is strangely at odds with the profuse means of communication that they have at their command. Indeed, some observers have seen in this process a human-made replica of older patterns of natural ignorance, whereby people today have become as unaware of the origin, the violence, the exploitation involved in the production of everyday articles and necessities as the peasantry was unaware of the forces that governed the rhythm of lives in bondage to the vagaries of the seasons and to the owners of the earth they cultivated. A new and artificial techno-peasantry is in the making: it is to this version of pauperizing people in the rich countries that the advertising industry is dedicated.[12]

If we think of the central thesis of the theory of "commodity fetishism" as consisting in the claim that the exchange economy masks the source of value creation in human laboring, then there are obvious parallels between it and Seabrook's arguments about the ingenuity with which the global market cultivates a modern consumer ignorance of the productive context of the commodity. Where the emphases differ, it is not so much because of any major theoretical incompatibility but in virtue of the specific conditions of modern capitalist production on which Seabrook is focusing—in particular, the relegation of so much of the Dickensian world of suffering and exploitation to the Third World "perimeter." The "veil of ignorance" about the source of the commodity is today that much easier to draw because the prepackaged item on the supermarket shelf is often enough remote by several continents from the misery and pollution of its production, not merely by the distance between high street and local factory. Or as Seabrook himself puts it, "The separation of consumers from producers has been one of the greatest triumphs of the global capitalist market. . . . The de-industrialization of Britain has been accompanied by a loss both of consciousness and memory."[13] And the further effect of this separation, of course, is that it obscures not only the source of value of the commodity, but also the environmental damage that so often accompanies its production.

OF FISH AND MEN

Underlying these Marxist arguments on alienation and fetishism there is an ontological thesis of particular relevance to current green debates. This is to the effect that the human species is self-creating through its productive and laboring activity. Marx depicts this as a process simultaneously involving a "humanization" of nature and a "naturalization" of the human. By the technologies whereby we extract and utilize resources for the satisfaction of human requirements, we "transform" the natural world, and the environment itself thus comes to bear the imprint (whether sealed in the pattern of a field of grain or the cement of an airstrip) of our particular patterning of need. At the same time, since we are creatures dependent upon an objective environment for becoming the subjects that we are, we thereby create our own "nature": we are returned to ourselves, so to speak, through the objective products of our industry, since these provide the context for all our aesthetic, moral, and cognitive experiences. This in turn implies that such experience must itself be viewed as the outcome of dialectical mediation: our subjective needs and senses (of sight, taste, etc.) acquire their objective existence in the products we create for their satisfaction, and these then condition our subjective sensibility and mold our future needs, aesthetic sense, and so forth.

Marx regards this "ontology of labor" as distinctive to the human species: humanity, he claims, differs from other animal species in producing in the form of an objective accumulation of cultural artifacts, languages, knowledges, institutions, and the like. This stock preexists the individual and is determining upon social life in general, but acts as an external matrix of behavior rather than in the manner of a biological inheritance. It is this contrast between the objective store of skills and knowledges available to humankind, and the restriction of other animals to those aptitudes that are genetically transmitted, that explains the ever-widening gap between animal and human capacities (whether to do good or ill), and it is associated in the Marxist argument with the possession of reflexive consciousness, and this in turn with our capacity to value:

> Man freely confronts his product. An animal forms objects only in accordance with the standard and the need of the species to which it belongs, while man knows how to produce in accordance with the standard of every species and knows how to apply everywhere the highest standard to the object. Man therefore also forms objects in accordance with the laws of beauty.[14]

Now this last is obviously an anthropocentric and somewhat dubious claim. It is not at all clear that man *does* know how to produce "in accordance with the standard of every species," or even what that phrase might mean. If it means, for example, that we can produce the honeycomb or the spider's web, then it is simply untrue—though we might, perhaps, be able to produce to bee or spider standards in the sense of making artificial replicas of combs or webs.[15] Nor is it clear what Marx means by saying that man knows "how to apply everywhere the highest standard to the object," given that the one thing that "man" seems everywhere given to quarreling about is what constitutes the "highest standard."

That said, Marx is surely right to direct attention to *valuing*, and the associated symbolic and aesthetic capacities, as distinctive to humanity, since it would seem true that we alone of organic nature are able to stand back to applaud or condemn the effects of our activities on the rest of nature. Indeed, ecological or any other form of social critique only makes sense in the light of the acceptance of our distinguishing powers of self-evaluation and self-change.[16] The ecologists do not address themselves to the AIDS virus, the muskrat, the Dutch elm beetle, the garden weed, or any other of nature's seemingly less ecofriendly species, precisely because it is assumed they lack the capacity to mend their ways or even to perceive the evil of them.[17] (This, incidentally, is not to say that the human species will manage to mend its ways but only to point out that the potentiality for it to do so is presupposed by any critical attack on its existing practice.)

Marx is also surely right to direct attention, through his "ontology of labor," not only to the ways we change the rest of nature, but to the ways in which our own nature is thereby changed. This is an important dialectic to sustain against the opposing essentialisms of both crude Enlightenment and "deep ecology" outlooks. Neither the picture of a rational humanity locked in struggle with an essentially "hostile" nature, nor its inversion in the idea of an essentially benevolent and blameless nature victimized by an instrumental and self-aggrandizing humanity, is helpful to the green cause. For both pictures in their differing ways tend to an unscientific denial of natural and human capacities for renewal and readjustment—a denial that in the end must undermine the cardinal green point about the *limits* on the possibilities of adaptation. They also tend to abstract from the positive impact humanity has had on the natural environment, and vice versa, and from the role this interaction has played in developing the love and respect for nature that will be an essential stimulus to future checks on ecological degradation. But, finally, it is only the more dialectical approach of Marxism that captures the truth that human beings, like other species, are creatures of their environment in the sense that their structure of needs and affectivity—their overall sensibilities—are formed in relation to their social and material situation and bear its imprint. Overcrowded, car-dense, high-rise, noisy, concrete-ridden environments have long been related empirically to crime, depression, drug abuse, apathy, and low expectations for life. The point I wish to make here is that the need for improved environments is stimulated in their provision. Understanding this dialectic, whereby in changing our policies on the use of nature we also enrich ourselves, will be crucial to mobilizing the political forces for any green revolution.[18]

I have suggested that we should avoid the essentialist picture of the humanity–nature relationship, in part because it fails sufficiently to acknowledge the capacity for ecological adjustment. On the other hand, there is also a risk that in stressing the dialectical quality of the humanity–nature interaction, one will understate the limits on change. Felix Guattari has recently drawn our attention to the octopus from the port of Marseilles that instantly collapsed and died on being placed in unpolluted water.[19] But should we conclude from this demonstration that all is well with the Mediterranean, or even with the quality of octopus life therein? No indeed. For not all species are likely to prove so accommodating, least of all those—like human bathers—who do not live in its waters but rely on them for less essential purposes. And as for the octopus, who can say what kind of a life this is for it, but one might speculate that it was nearer to that of the drug addict or the alcoholic than to that of the healthy individual. In any case, even if most species were to prove as adaptive as the octopus— which is far from being the case—the main limits on ecological viability would still remain. For these include physical limits on natural resources— such as fossil fuels and minerals—that of their nature cannot "adjust" to their growing scarcity and are, quite simply, nonrenewable and irreplaceable.[20] Or else they derive from the physical–chemical properties of things in conjunction with the physiology of organic species, including our own. The biology of humans, seals, salmon, and so on is such that there is no withstanding certain forms of toxicity (most of them humanly produced) such as radiation, chemical poisons, oil slicks, and the like. In fact, the whole history of planetary life suggests that adaptation to these kinds of poisons and pollutions is simply out of the question, so that any appeal to nature's wondrous powers of self-renewal is grossly misleading as to the true nature of our current crisis.

Now if we ask at this point where Marx stands in regard to all this, then I think the answer is that he stands ambiguously with a foot in both camps, but mainly weighted, I shall argue, on the side of sustaining a humanity–nature dialectic that is also cognizant of limits on nature. For example, if we look at the closest he comes to commenting on the octopus, he certainly seems biased in this direction. The proletariat, he tells Feuerbach in the *German Ideology,* will take as kindly to the idea that its "essence" is realized in the living conditions of its capitalist "existence" as the fish to the idea that its "essence" is equally realized whether it swims in clean or polluted water.[21] Against Feuerbach's abstract essentialism (which finds the "essence" of every species everywhere realized in interaction with its conditions of existence), Marx is here pointing out that the real, as opposed to the "philosophical," essence of things is not so elastic: that just as "man" is incapable of realizing his essence or truth in capitalist conditions, so fish nature is such as to rule out any realization of its essence in the port of Marseilles. The emphasis here is clearly on the limits imposed by the intrinsic nature of any entity or species on the forms in which the potentialities inherent in that nature can be actualized.[22] It is of the nature of the tulip bulb to require an uncontaminated soil, a sufficiency of warmth and water, and much more if it is ever to actualize its potentiality to become a fully grown, blossoming plant; and so on.

OF LIMITS, PRESUPPOSITIONS, AND NATURE

It is true that in the later discussion of alienation in the *Grundrisse* we are offered an account of human development that is seemingly less congruent with the recognition of natural limits implied in the earlier argument around "species-being." Alienation in the *Grundrisse* is presented less in terms of the loss or lack of actualization of an essential species-being under capitalism, and more in terms of the latter's destruction of all natural ties relating the individual to a given environment, and providing him or her with what Marx describes as a "direct extension of self in the inorganic." In virtue of the impersonality and generality of capitalist society (where the worker figures as an exchangeable unit of laboring capacity rather than as occupant of a particular role) the individual is deprived of the "objective presuppositions" of selfhood that come from being tied to a specific place and community: the worker under capitalism, says Marx, is "objectless" and "naked in his subjectivity."[23] But Marx also insists that this same process is that which frees the worker of all "presuppositions" and bonds with the "inorganic conditions of life," and thus, in principle, places the individual in a position to break out from any limited and predefined selfhood. Pure subjectivity is here conceived as the condition of escape into any and every possible mode of objectification. Capitalist production may create "object-lessness," but it also drives labor "beyond the limits of natural paltriness"; and whereas previous stages of production represented "mere local developments of humanity" and "nature idolatry," under the spur of capitalist industry,

> [Nature] ceases to be recognized as a power for itself and the theoretical discovery of its autonomous laws appears merely as a ruse to subjugate it under human needs, whether as an object of consumption or as a means of production. In accordance with this tendency, capital drives beyond the national barriers and prejudices as much as beyond nature worship, as well as all traditional, confined, complacent, encrusted satisfactions of present needs, and reproductions of old ways of life.[24]

And associated with this line of argument goes the well-known salute to the "universality of individual needs, capacities, pleasures, productive forces, etc." that are to be made available through universal exchange once "the limited bourgeois form" has been stripped away and the human individual no longer "strives to remain something he has become, but is in the absolute movement of becoming."[25]

Against, then, the recognition of limits on human fulfillment and gratification implicit in the *German Ideology*'s approach to the realization of the essence of species-being, we must set these more Promaethean aspirations to break loose from all natural limitations and essential presuppositions into a utopia of hedonistic insatiability. Nonetheless, Marx is aware of the tensions in the Promethean position:

> In bourgeois economics—and in the epoch of production to which it corresponds—this complete working-out of the human content appears as a complete emptying-out, this universal objectification as total alienation, and the tearing down of all limited, one-sided aims as a sacrifice of the human end-in-itself to an entirely external end.

> This is why the childish world of antiquity *appears on the one side as loftier*. On the other side, *it really is loftier* in all matters where closed shapes, forms and given limits are sought for. It is satisfaction from a limited standpoint; while the modern gives no satisfaction; or where it appears satisfied with itself, it is vulgar.[26]

He is very aware, too, I suggest, of exactly how much blame must be attached precisely to the "limited bourgeois form," in particular to the subordination of bourgeois society to the quest for profit, for diverting the potentialities opened up by modern industry into rapacious and destructive channels. Indeed, in several pronouncements, it is precisely the *contrariness* of the "bourgeois form" from the point of view of human and natural well-being that is emphasized. Thus, for example, having compared progress in exploiting societies to that "hideous pagan idol who would not drink the nectar but from the skulls of the slain," he goes on:

> In our days everything seems pregnant with its contrary. Machinery gifted with the wonderful power of shortening and fructifying human labor, we behold starving and over-working it. The new-fangled sources of wealth by some strange weird spell are turned into sources of want, the victories of art seem bought by the loss of character. At the same pace that mankind masters nature, man seems to become enslaved to other men or to his own infamy. Even the pure light of science seems unable to shine but on the dark background of ignorance. All our inventions and progress seem to result in endowing material forces with intellectual life, and in stultifying human life into a material force.[27]

Admittedly, this passage speaks of "mastering" nature and is clearly committed to the idea that technology and science are in principle benevolent forces: Marx's position is in no sense nostalgic, Luddite, or irrationalist. But that in itself need not make it ungreen. One may argue, on the contrary, I think, that an ecological and socialist corrective to the depredations of the "bourgeois form" of wealth production will depend in part on a highly rational, technically sophisticated intervention in ("mastering" of?) natural forces. To correct soil erosion (whether resulting from human or purely natural causes) we cannot leave nature to itself. We will need positively to do some things (plant trees, build dikes, etc.) and to stop doing others. And one may cite many other examples where a "hands-off nature" approach would have ecologically deleterious consequences.

However, there is a further dimension to Marx's emphasis on the "negatives" ensuing from the *bourgeois form* of harnessing resources, which is of cardinal pertinence to the green critique of industrialism. This lies in the clarity with which, by exposing this form *as a form* (i.e., as in no sense a necessary mode of production) he also exposes the nonnecessity of crystallizing all surplus labor in commodities.

Since profit can only be realized through the sale of commodities (material goods and services almost always involving material resources), any society subject to capital accumulation will be driven into a wholly antiecological concretization of surplus labor time (time, we might say, left over after producing the necessities of a decent and modest life-style) into surplus product—into proliferating luxury production. But as Marx over and over again makes clear, a socialist economy can enjoy surplus labor time as *free* time: as time not spent on producing, and therefore a fortiori not expended in resource-consuming ways. Or to put this less tendentiously, he makes clear that it is not part of the very nature of wealth production that all available social labor time should unceasingly become embodied in new material commodities and services. He makes clear that one could in principle opt for a more reproductive system of "primary" need satisfaction.

It is true, of course, that what counts as "primary" or "necessary" consumption as opposed to "affluence" and "luxury" consumption is not fixed by nature but is very much a convention varying for different times and places. All people at all times have a need for water. But the clean water piped into one's home that even the more ecologically ascetic of the Greens would seem to regard as part of necessary consumption in the industrialized nations remains a luxury relative to what counts as "primary" need or necessary consumption in other parts of the globe. In other words, a "politics of need" is presupposed by any attempt to distinguish for any actual society between the part of social labor time that is "necessary" because it is essential to the satisfaction of a certain range of goods deemed necessary, and the part that is "surplus" (and that could be spent either in idleness or in creative/productive activities to satisfy a range of more sophisticated needs). What is needed even in a basic or primary sense is culturally conditioned to a very high degree, albeit determined in an abstract and general sense by our common natural physiology as a species.

It is also true that when Marx analyzes necessary and surplus labor time under capitalism and its implications for the release of "surplus" or "free" time in a socialist society, he does not actually specify that such surplus time would in fact be spent either literally doing nothing or only in ways that would be sustainable in terms of resource use. His argument in this sense is not explicitly ecological. In other words, there is nothing overt in Marx's argument that associates socialism with a restraint on material and resource-intensive forms of production or with a use of surplus/free time that would be obedient to environmental limitations on the expansion of certain forms of consumption. In this sense, a socialist economy (whether organized in more or less centralized ways) does not in itself guarantee that surplus time will not be devoted to expanding the range of material commodities available to society. And if such a society does opt for this use of its energies, then it will not prove much of a corrective to the "bourgeois form" from any ecological perspective.

But the important point is that it is only under a socialist economy that a society is placed in a position to choose the forms of embodiment of its labor our time and therefore in principle is able to opt for a more ecologically sustainable pattern of consumption.[28] It may be true that Marx's dialectic of necessary and surplus time under socialism leaves open the question of what is "needed," but he certainly exposed more clearly than any other theorist the ways in which surplus time *could* be spent not in enhancing "living standards" in the conventional sense of the term, but as "free" time: as time spent not producing material goods and therefore not expended in resource-consuming ways. He reveals socialism as the possibility of idleness; and in this idleness, one may argue, lies one of the most important ecofriendly resources available to human societies at the present time. For in the last analysis, it is only if we stop working, in the sense of devoting labor time to the production of resource-hungry material commodities, particularly in the more affluent global regions, that we shall stave off the barbarism of ecological collapse.

This is also, in the last analysis, why a commitment to a Marxian-socialist economic approach, which insists on the necessity of correcting the inequalities and ecodestructiveness of capitalism from the *production* rather than the *distribution* end, should be a sine qua non of anyone who is seriously anxious about the future ecological survival of the planet. Redistributing income through heavier taxation on the rich, basic income schemes, and similar policies, will correct some of the economic injustices and will help shift investment away from luxury production. But in the first place (as Marx points out in his criticism of Proudhon and the English "Ricardian Socialists"[29]) full equality in the distribution of wealth is incompatible with the maintenance of the competitive market economy; and, in the second place, any redistribution of wealth that is compatible with the continuation of capitalist relations of production will do little to check the expansion of material commodity production. A limited social democratic reallocation from rich to poor cannot in itself halt and reverse the (material) commodifying and (material) consumerist tendencies that are the real enemies of ecological revolution.

I have suggested that in a formal and abstract sense, Marx's argument on the relations between necessary and surplus labor time leaves open the question of the actual forms consumption would take in a socialist society, and therefore does not explicitly commit the latter to an ecologically responsible use of labor time. But there is little doubt, all the same, that Marx himself envisaged a world in which freedom from the "laws" of capitalist accumulation would lead to the emergence of a very different structure of needs: a structure wherein the major source of satisfaction would be the free time released for the cultivation of "rich individuality" conceived in essentially spiritual and artistic terms. For whatever Marx's more technocratist followers may have advocated as the consumption goals of a "Marxist" economy, no one who reads Marx's works intelligently and sympathetically can really claim that his is a consumerist perspective. On the contrary, there is even, if one dare say so, something a little old-fashioned and otherworldly about the picture Marx conjures up of the many-sided "hunter, fisher and critical critic" of the communist future.[30] At any rate, the idea he conveys of a "rich development of individuality" is hardly rampantly materialist, even if the hunting and fishing (presumably in unpolluted waters) will not recommend itself to all green sensibilities.

Finally, it should be recognized that there is an important ambiguity attaching even to Marx's more Promethean and seemingly least green pronouncements. When he refers us, for example, to a communist future "unmeasurable by all previous yardsticks," and associates this future with the escape from "natural limits" and "presuppositions," we are inclined to view this as a piece of heady Enlightenment optimism very much at odds with ecological cautions about dwindling resources and the need for a sustainable level of consumption. But there is no reason to interpret the idea of "breaking with natural limits" primarily—let alone exclusively—in terms of breaking with limits on natural resources; and there is very little, in fact, in Marx's argument that invites us to think of it in that way. On the contrary, the bias of the argument is more toward the idea of transcending a previously self-satisfied, but limited, human nature, and the gauges of progress and human well-being that have gone together with that "nature." If we construe it in this sense, then the argument is by no means so incompatible with ecological demands, and might even be said to be quite consistent with the green call for a "new philosophy" in our approach to nature. For it would have to be admitted, I think, that if we were indeed to break with "human nature" as hitherto manifested in our treatment of the rest of nature, we would be breaking with attitudes that had issued in some very wasteful and destructive patterns of use of our environment. Likewise, in respect of previous yardsticks of progress: to move into a future "unmeasurable" by these standards would be to move into a future that had broken with some profoundly antiecological conceptions of the "good life," "civilization," and human "development."

I would stress here that I am drawing attention to an *ambiguity* in Marx's "Promethean" discourse, and not trying to claim that the "ecofriendly" construction I have put upon it is that which Marx himself intended and which we should accept as "faithful" to Marxism. Rather, the point is that the ambiguity in Marx's position derives from the same vacuity or lack of content in the projection of a future beyond "all previous yardsticks" that makes it impossible to claim any one interpretation is definitively Marxist. Where the vision of the future is so underdefined, nothing definite can be said of it. In other words, statements by Marx that have been received as anthropocentric and arrogant in their approach to nature are often so formally dialectical and devoid of empirical content that they allow for either a more green or a more technocratist interpretation. This is one reason for wanting a synthesis of green and Marxist intellectual forces: it will allow the ecological argument to flood the vacuum. Where Marx failed to blueprint, we can blueprint green.

NOTES

1. "The green movement," writes Peter Tatchell (see "Ecological Sustainability" in *Into the Twentieth Century,* ed. Felix Dodds [Basingstoke, U.K.: Green Print 1988], 38) "is the most important new radical movement since the emergence of socialism nearly two centuries ago." Cf. Dodds himself (in the same source): "Just as socialism became the dominant philosophy of this century in reaction to the existing political climate, green politics could well become the philosophy of the twenty-first century" (xi); also see Jonathan Porritt (same source): "This multiplicity represents the single most important social and political movement since the birth of socialism" (196).

2. There are many voices here, among whom one might single out Rousseau, Wordsworth, and William Morris. See Keith Thomas, *Man and the Natural World: Changing Attitudes in England, 1500–1800* (London: Allen Lane, 1983); and also see John Passmore, *Man's Responsibility for Nature* (London: Duckworth, 1980), Part 1.

3. For statistics on pollution in Central and Eastern Europe, see the essays by Mark Thompson, Kate Soper, and Martin Ryle in *Something in the Wind: Politics after Chernobyl,* ed. L. Mackay and M. Thompson (London: Zwan, 1988); also see the essays by Istvan Rev, Kate Soper, and Martin Ryle in *The New Detente,* ed. M. Kaidor, G. Holden, and R. Falk (London: Verso, 1989).

4. The reorganization of national economies along socialist lines is a precondition of any green revolution because it allows for a political control over the use of resources—a control denied to capitalist economies, which in the end are answerable only to the logic of value accumulation. But being in a position to exercise such control does not in itself decide how it will be used, and if consumerist economic goals are chosen there will be an ecological price to pay. See the concluding pages of this chapter.

5. Such as has been given, for example, by Gerry Cohen in his *Karl Marx's Theory of History: A Defence* (Oxford: Basil Blackwell, 1978), a work that relies very heavily indeed on the argument of the 1859 "Preface."

6. R. Bahro, *Socialism and Survival* (London: Heretic, 1982), 27; cf. Martin Ryle, *Ecology and Socialism* (London: Radius, 1988), 68–70.

7. Jonathan Porritt, for example, in his "Preface to the Programme of *Die Grünen*" (English translation in *Seeing Green* [Oxford: Basil Blackwell, 1984]) hails the end of "the redundant polemic of class warfare and the mythical immutability of the left/right divide" (4), and also argues at another point that "genuine redistribution of power can no longer be simplistically interpreted in terms of setting class against class" (226). In a general sense, as Martin Ryle argues (*Ecology and Socialism,* 20), environmental groups and green parties "tend to see themselves as . . . expressing a 'general interest'–the interest of nature, of a viable humanity/nature relationship–that is distinct from and even in opposition to the interests of each and every particular group/class within a social formation."

8. Cited in Ryle, *Ecology and Socialism,* 33.

9. In that Hegel defines alienation in terms of the failure of Spirit to realize its own conceptualizing role in what it takes to be "objective." For a full discussion of the differing but comparable usage of the term by Hegel, Feuerbach, and Marx, see Chris Arthur, *The Dialectics of Labour* (Oxford: Basil Blackwell, 1987).

10. These are my own examples, but the essential idea is central to green thinking from William Morris onward. For recent examples, see the opening and closing essays of Jeremy Seabrook, in *The Race for Riches* (Basingstoke, U.K.: Green Line, 1988); Ryle, *Ecology and Socialism,* 27–31, 43–58, 75–78; and William Ophuls, *Ecology and the Politics of Scarcity* (San Francisco: W. H. Freeman, 1977), esp. 167–183.

11. Karl Marx, *Collected Works* (London: Lawrence and Wishart, 1975), 3:326.

12. Seabrook, *Race for Riches,* 101.

13. Ibid., 102–103.

14. Marx, *Collected Works,* 3:277.

15. There is a further question, of course, as to whether we could produce such replicas in a manner allowing the insects concerned actually to make use of them.

16. In an interesting article on Marx, "Humanism vs. Speciesism" (*Radical Philosophy, 50,* Autumn, 1988, 4–18), Ted Benton takes issue with Marx's human–animal dualism on the grounds that the general qualities attributed exclusively to humanity by Marx are a property of other animal species too. He proposes instead a naturalistic approach deriving from the central insight that "these things which only humans can do are generally to be understood as rooted in the specifically human ways of doing things which other animals do" (14). The capacity to value, which I here associate with Marx's human–animal dualism, is not directly discussed in the article, but Benton suggests that the starting point of any analysis of the distinguishing "aesthetic, cognitive, normative, 'spiritual'–in other words 'cultural'–dimension to the way in which humans meet their physical needs" is the

> recognition of a need which is common to both humans and non-human animals. The specification of the distinctively human then proceeds not by identifying a further, supervenient class of needs possessed only by humans, but rather by identifying the species-specific way in which humans meet the needs they share with other species. (15)

But the weak point of this approach, it would seem, is that it cannot explain *why* there is a need for human beings to do things in their more specifically aesthetic ways—*why,* for example, they don't just eat more as animals do. And this weakness emerges, I think, rather clearly in Benton's response to the objection that not all human needs would seem reducible to physical need, since he offers us only the idea of viewing them "as *in some sense* consequential upon those needs which are common to natural beings" (15), when, of course, it is precisely the particular sense that is here in question. For the naturalistic, antidualist position to prove compelling, it will need to do more than recognize a human distinctiveness in the ways of doing things other animals do: it will need to offer a cogent naturalistic explanation for the difference itself.

17. I understand, however, that some "deep ecologists" would regard any destruction of nature's more pestilential aspects as an illegitimate interference in the global ecobalance—and would include even the attempt to develop anti-AIDS drugs and vaccine in this condemnation.

18. My remarks here should not be taken to imply that Marx fully recognized, let alone conceptualized, these implications of his account of the humanity–nature interaction. For some pertinent indications of the conceptual inadequacies of Marx's account as it sounds, see Ted Benton's contribution to this volume (Chapter 8)—in particular, his criticisms of Marx's emphasis on the "transformative" and "intentional" aspects of the labor process at the expense of more "ecoregulatory" types of practice in his section The Labor Process: Ecoregulation (pp. 160–163).

19. Felix Guattari, "Les Trois Ecologies" (Paper delivered at the Institute of Contemporary Arts, London, November 1988).

20. There is considerable debate, obviously, about the extent to which we are likely to be able to provide substitutes for exhausted or dwindling resources, but there is general agreement (reflected in such authoritative sources as the Brundtland report, *Our Common Future* [1987] that it is out of the question that alternative forms of energy could be developed sufficient to bring global use generally up to the level of the currently most affluent nations—even if it were desirable or feasible to do so in other terms). Cf. Martin Ryle, *Ecology,* 3–4. (Any pronouncement on the potential of the recent nuclear fusion experiments would be premature.)

21. Marx, *Collected Works,* 3:58–59.

22. Readers will recognize that such arguments rest on attributing an Aristotelian notion of "essence" to Marx along the lines most systematically defended by Scott Meikle in his *Essentialism in the Thought of Karl Marx* (London: Duckworth, 1985). I think, in fact, that there are considerable problems about treating Marx's approach in his later work to human self-realization along such teleological lines (and some of these difficulties are obliquely touched on in my discussion below on the approach of the *Grundrisse* to alienation). But the more Aristotelian approach does indeed seem justified by the earlier argument around the notion of "species being."

23. Marx, *Grundrisse* (Harmondsworth, U.K.: Penguin, 1973), 450–456; and also see the whole section on precapitalist economic modes of production, 471–515, where Marx elaborates on this "theory of personality." See also my own

discussion of this topic in *On Human Needs* (Brighton, U.K.: Harvester Press, 1981), 125–142.

24. Marx, *Grundrisse,* 409–410.

25. Ibid., 487–488.

26. Ibid.

27. Marx, "Speech at the Anniversary of the 'People's Paper,' " in *Selected Works* (London: Lawrence and Wishart, 1972), 1:500.

28. But this level of consumption would not necessarily be without variety or devoid altogether of luxury goods. André Gorz's "utopia," for example, would include a sphere providing for "wants" in addition to that satisfying "basic needs." See his *Paths to Paradise* (London: Pluto, 1985), and his *Farewell to the Working Class,* (London: Pluto, 1982); and see also Martin Ryle, *Ecology,* 75–78.

29. For Marx's strictures against Proudhon, see *The Poverty of Philosophy,* in *Collected Works,* 6:120–143; and on the English "Ricardian Socialists" (Bray, William Thompson, John Gray, Thomas Hodgkin, Thomas Rowe Edmonds, etc.), see 143–150; also see *Grundrisse,* 319, and *Theories of Surplus Value* (London: Lawrence and Wishart, 1972), Part 3, 319–325.

30. As a jibe against Stirner, this is not a phrase to be taken very seriously as a guide to Marx's true views of "communist man," and I hope readers will appreciate that I use it here myself with a touch of the same irony. But the essential point I think holds good, and Marx was certainly more of a Millian gentleman than a Benthamite sausage maker in his estimations of the quality of human happiness.

Toward an Ecological Marxism

INTRODUCTION
TO PART II

This part of the book brings together a small selection from a quickly growing literature that seeks to build from a critical revision of the Marxist classics towards an ecologically informed historical materialism. This selection begins with an important work of historical recovery: Arran Gare's revelatory exploration of the early history of Soviet environmentalisms. As Enzensberger's and Soper's contributions, in Part I, acknowledge, the disastrous legacy of environmental destruction in the Soviet Union and Eastern Europe remains a profound obstacle to any red–green realignment. Gare's study, which draws upon newly available sources, and in particular upon the monumental work of Douglas R. Weiner, shows that the tragedy goes even deeper than this. Between the Revolution of 1917 and the full consolidation of Stalinist power in the early 1930s, there flourished in the Soviet Union a diverse environmental movement, while Soviet ecologists led the world in the development of their science, and attempts were made to put Soviet agricultural and economic development on a sound ecological basis. Stalin's victory over these forces set the Soviet Union on course for an ecologically and socially disastrous programme of centrally directed, forced industrialization. The science of ecology was "virtually suspended" for twenty years. Soviet science was put at the service of the drive for industrialization, paving the way for the doctrines of "proletarian science" and the notorious Lysenko episode. This, too, was a disaster both for the study of biological inheritance, and for Soviet agriculture.

So, there was no inner necessity in the subsequent antiecological history of the Soviet Union. As Gare shows quite convincingly, the flourishing of ecological perspectives in the 1920s was facilitated by the revolution, while Stalin's reversal of these developments involved a revival of prerevolutionary Russian nihilism. Sadly, not only was this period of Soviet history concealed from the Soviet people themselves, but its developments in linking ecology to socialist thought and practice remained undiscovered by later "Western" Marxists when they finally came to address these very problems from the 1960s onwards.

The remaining three selections in Part II represent but a small fragment of the creative work that has gone on since then. The debate has been an international one, with many contributions from Australasia to the Indian subcontinent, from North and South America, as well as from Western Europe. Advances have been made in four broad areas of work. First, the concepts of historical materialism have been revised and developed, with particular attention paid to the Marxian theory of the capitalist mode of production, so as to offer an *explanatory account* of the production of ecological degradation and crisis. Second, the socialist project has been redefined to give due weight to its ecological dimensions. This second task turns out to pose basic questions of a normative kind, some of which remain at issue between radical Greens and many ecosocialists. A third area of work, less developed so far than the others, concerns the development of a Marxian or socialist normative framework in environmental philosophy. Finally, any redefinition of the socialist project urgently calls to attention questions of agency and strategy. How should the currently existing environmental social movements, organizations, and parties be understood? What are the prospects for a realignment of left and green politics? What might be the relationships between any such realignment and other autonomous social movements such as feminism, antiracism, gay and lesbian politics, and so on?

Gunnar Skirbekk's pioneering essay, written in the wake of the oil crisis of 1974, clearly acknowledges the extent of the rethinking of Marxism that ecology makes necessary. Skirbekk argues for two crucial revisions in the Marxist theory of capitalist production if ecological crises and their consequences are to be comprehended. The first revision, which anticipates James O'Connor's thesis concerning the "second contradiction of capitalism," is that capitalism is, indeed, moving toward a self-destructive crisis. However, this is not the crisis resulting from the "opposition" between forces and relations of production postulated in Marx's original theory. Rather, it is an "ecocrisis" resulting from opposition between the forces of production and the natural "conditions of production," which Skirbekk also refers to as an opposition between the forces of production and the forces of nature. A clear requirement for theorizing this opposition is that the traditional Marxist concept of the "infrastructure" in any social formation has to be revised to include not only forces and relations, but also natural conditions of production. Skirbekk's second revision involves a distinction between what he calls "reproductive production" and "extractive production." In the case of the latter, surplus profits are made possible by not only discounting the costs of current ecological deterioration, but also by depriving future generations of the use of resources currently available. In these industries, therefore, it is not only labor that is exploited, but also nature and, indirectly, future generations.

Though his essay predates the enormous "growth industry" of environmental economics, Skirbekk provides a concise and cogent critique of its central strategy: conserving scarce resources and "internalizing" environmental externalities by including them in commodity prices through taxation or regulation. In Skirbekk's view, this strategy is ultimately irreconcilable with capitalist accumulation because it threatens profitability, competition, and/or the reproduction of labor power. Ecologically sound production cannot, at the same time, be capitalist production. He also makes short work of the idea that moral pressure on and from consumers might suffice to resolve ecological problems on a capitalist basis. Clearly, both these environmental reform strategies have developed apace since Skirbekk's article was written, and the issue of how far they might work, and with what consequences, remains very much open to debate. Nevertheless, Skirbekk's arguments, especially his point about the incommensurability of environmental and other goods, remain pertinent.

On the question of redefining socialism, Skirbekk has relatively little to say. Nevertheless, his brief indications are worth developing. First, it follows from the inclusion of the natural conditions of production within the concept of the infrastructure that socialism as a mode of production must now be conceptualized as entailing a reconciliation not just between forces and relations of production, but also between these and the natural conditions of production, or "forces of nature." This requires, as in the traditional conception of socialism, a planned, or political, direction of economic activity, but as the example of the Soviet Union shows, this is no guarantor of ecological soundness. What will be needed is a shift in the balance between industry and agriculture, and a shift from "extractive" to "productive" or "reproductive" activity.

These shifts, in turn, will require a very significant revision in the value system of "actually existing" socialist societies along two related dimensions. First, there will be a need to foster respectful, in contrast to aggressive and exploitative, attitudes toward the laws of nature. Second, human fulfillment will itself have to be understood less in terms of technologically driven "progress" to higher material living standards, and more in terms of "cultural, social, and political values." Like some thinkers in the anarchist tradition of social ecology, Skirbekk refers us back to the visions of the "good life" advocated by the classical Greek philosophers.[1]

Finally, on agency and strategy, Skirbekk remains resolutely committed to a class analysis, but undercuts this by way of a powerful demonstration of the strategic emptiness of a concept of the working class defined solely in terms of economic exploitation. There are immense disparities between the material conditions of life for First and Third World workers, and the impact of ecological crisis is likely to be felt unevenly and differentially between different groups of workers. This suggests two things. First, that a fascistic, racist, and authoritarian response to ecological crisis cannot be ruled out: socialism is by no means the inevitable outcome of the ecological crisis of capitalism. Second, that the whole Marxist corpus of concepts for thinking about strategy and transition, including the concept of class itself, needs to be rethought in the face of the contradiction between capitalist production and the forces of nature.

With *Green Production*, which is in the same series as this book, Enrique Leff's work has been presented. However, his approach has such distinctive features that it is quite essential to have it included in the present collection. Though Leff's language is often hard to follow, the rich theoretical content and profound practical implications of his work do repay the effort. Over twenty years, Leff has been working on a theoretical project of massive scope: to overcome the theoretical and epistemological problems of an integration of biological, geographical, and sociohistorical disciplines; on this basis to construct a synthesis out of historical materialism and ecology; and then to work out the implications for social and ecological transformation and "ecodevelopment."

Leff's starting point is an environmentalism firmly situated in the context of Third World poverty and ecological destruction This is the key to understanding his rejection of "zero economic growth" and his advocacy of a new productive paradigm. The commitment to development must not be abandoned, but a new paradigm of development must be constructed that combines enhanced consumption with ecological sustainability, social equality, democratic participation, decentralization, and cultural pluralism.

Despite his often negative evaluation of other work seeking to unite Marxism and ecology (readers may wish to make up their own minds, e.g., about the justice of his dismissive approach to Enzensberger's work), he does share much with the other writers presented in this collection. In particular, like them, he sees capitalist production relations and capital accumulation on a global scale as the primary source of environmental destruction. Moreover, in considering the role of combined and uneven development in both the destruction of conditions of production and in the failure to utilize natural resources, Leff can be seen as closer to the classical Marxian approach. For him, too, capitalist relations are a fetter on the development of productive forces, which would be liberated in a postcapitalist future society.

However, the approach is much more radically innovative than this might suggest, since Leff's work transforms the classical Marxian concept of "forces of production" and so gives a radically new content to their "development." The two key features of his transformation are, first, the inclusion of ecological processes among the forces of production, and, second, the recognition of cultural resources as forces of production. The primary productivity of natural processes, ecosystems as "productive potential," should be subjected to a form of integrated, participatory management that selectively enhances their provision of use values.

This cuts against any view of ecodevelopment as a set of universalizing remedies, imposed from above, and centered on the conditions for the production and reproduction of capital. The new productive rationality that Leff proposes starts explicitly from the aims and requirements of local communities, and draws upon local material cultures which have themselves developed historically in relation to the specificity of local ecosystems. Technologies developed elsewhere could be incorporated into any such ecodevelopment strategy, but only on the basis of their appropriateness to local ecological and cultural conditions and processes.

Though this proposal is recognizably a socialist one, it does not seek to prescribe any specific model of development. Rather, the intention is to enable a proliferation of qualitatively different development paths, in accordance with differences in local communities and their environments. So, compared with more traditional Marxian views of agency and strategy, there are some clear departures. Moral and political values attributed to environmentalists and other "new social movements," such as welcome for cultural diversity and multiple identities, the rejection of centralized power, and universalizing projects, are endorsed by Leff. But there is, at the same time, a recognition that class interests are at stake, and that some form of unified political action will be needed to shift existing power relations. How far these strategic requirements can be consistently combined with the normative framework of decentralism and pluralism is yet to be explored and tested in practice.

Not only does Leff provide us with a rich and challenging approach to the key issues of Third World poverty within the context of environmental socialism, but he also demonstrates the indispensibility of historical materialism to this question. The concept of qualitatively different modes of production as specific combinations of cultural, technical, and ecological resources and relations enables us to see the question of "ecological limits" as both real and important, but at the same time as one that is *relative* to each specific mode of appropriation of nature. The implication of this is a view of the future as an open space of bounded possibilities, rather than as an increasingly self-destructive unilinear "development."

Leff's other important contribution is the avoidance of a certain "economism" in the concept of the economy itself, through his introduction of the idea of cultural resources as forces of production. One potential problem with this, however, is that it seems to leave untheorized the broader dimensions of culture insofar as it is not involved in production. Though his text includes, for example, a recognition of the aesthetic value of nature, and also of issues to do with "the quality of life," it is unclear how much space is left in a social formation animated by the new productive rationality for more recreative, contemplative and aesthetic forms of relationship to environments. In the words of Vaillancourt's title, Leff's environmentalism is, like Marx's, more Benedictine than Franciscan—appropriately enough given the depth of the problems of distorted development, poverty, and ecological destruction that it addresses.

My own contribution runs parallel to Leff's in quite instructive ways, although the environmental ethic that informs it is somewhat different. The overall purpose of my argument is to explore and explain those features of Marx's economic theory of capital that have most obstructed the development of historical materialism as an explanatory theory of ecological crisis. On this basis, I propose a series of conceptual revisions, which in turn generate a further open-ended research programme. I argue that, if appropriately revised, the Marxian economic theory of capital offers a powerful explanation of the relation between capital accumulation and ecological crisis. In passing, I note some of the ways in which the proposed revisions to the concepts of labor process and mode of production might provide the basis for an account of the social distribution of environmental impacts. This, in turn, might offer a way of going beyond traditional class analysis in thinking about sources of collective agency for ecological socialist politics.

While it is recognized that there is much in the writings of Marx and Engels that shows sensitivity to environmental conditions, my emphasis is on key conceptual weaknesses in the economic theory of *Capital*. Having distinguished capitalist production as a process of production of use values by means of specific, "concrete" labor processes, from capitalist production as a process of self-expanding value, Marx was almost wholly concerned with developing his theory of the latter, together with the social–relational contradiction between capital and labor that was intrinsic to it. Because Marx saw the dynamic of the whole system, and the prospects for transition to socialism, as stemming from the contradictory character of the production and realization of value, his relative neglect of labor processes as specific forms of organization of labor and technology in the appropriation of natural objects, forces, and processes was understandable. Moreover, Marx's weakness in this respect was reinforced by his insufficiently radical critique of classical political economy, which largely shared these failings; by the residues of his political hostility to Malthusian "natural limits" arguments; and, finally, by his participation in the widespread nineteenth-century "Promethean" celebration of the transformative power of modern science and technology.

The central argument of the chapter is that in subordinating his account of the labor process to the requirements of his theory of value production and economic class conflict, Marx neglected to develop an account of labor processes as, simultaneously, forms of organization of human labor in relation to the appropriation of nature. Working from Marx's own starting point, I attempt to show, first, that Marx inappropriately assimilates quite different labor processes to a single, "productive–transformative" model, and, second, that even this "productive–transformative" model is conceptualized in ways that radically undertheorize the significance of both produced and naturally given conditions of production, and unintended (environmental) consequences of productive practices.

The concept that bears most of the weight of this argument is what I call the "intentional structure" of the labor process. It is in terms of this intentional structure that specific objects, mechanisms, substances, and the like can be identified as "instruments," "raw materials," "products," and so on. I argue that "ecoregulatory" practices such as farming, forestry, horticulture, and so on, as well as processes of primary appropriation such as fishing, mining, hunting, and the like, have quite different intentional structures from one another, and from productive–transformative labor processes such as handicraft production and industrial production. The assimilation of ecoregulatory and primary appropriative labor processes to productive-transformative ones has the effect of concealing the extent to which modern industrial capitalist economies continue to be constrained by the context-dependence, seasonal rhythms, geographical fixity, and relative nonmanipulability of many of the conditions of such activities as agriculture, mineral extraction, and energy generation. However, it is also the case that Marx characterizes the "intentional structure" of productive–transformative labor processes themselves in a way that radically undertheorizes their dependence upon naturally given and produced contextual conditions, upon human-reproductive labor processes of several kinds, and upon mechanisms, materials, and energy whose use has effects that exceed and sometimes undermine the purposes to which they are put.

This suggests a program of research that would correct Marx's account of "the" labor process by means of, first, a typology of the different intentional structures instantiated in particular working practices, and, second, a wider specification of each labor process in terms of the potentially ecologically significant *un*intended consequences that result from the combination of a particular intentional structure with its material embodiment: for example, the exhaust gases, chemical wastes, energy releases, and so on that are ecologically significant *effects* of labor processes, but that are not captured in any *intentional* characterization of the labor process in terms of its raw materials, instruments and product. This, in turn, would need to be recombined with Marx's analysis of the overriding *capitalist* intentional structure of surplus-value maximization to provide a three-dimensional characterization of any example of capitalist production: social relations of production, labor process as intentional structure, and labor process as material combination of causal mechanisms.

This approach can now be turned into a provisional strategy for explaining why capitalist production tends to undermine its own ecological conditions. The forms of economic calculation that govern patterns of capital investment and accumulation, and, through that, the distribution of activity as between different "concrete" labor processes, are conducted in terms of abstract values. They are thus not only highly insensitive to qualitative differences between different concrete labor processes and use values but also, and for the same reasons, insensitive to ecological context-dependence, ecologically significant unintended consequences, timescales of organic processes, and so on. Marx's undertheorization of these aspects of capitalist economic calculation merely reflects actually existing practice, the result of which is a tendency to *materially* assimilate all labor processes to the productive-transformative model, and to drive productive practices themselves beyond their sustainable limits.

However, there are important limits to the argument thus stated. What is characterized here is a general *tendency* of capitalist production. The consequences of any actualization of this tendency in any specific branch of production could not be predicted in advance of the kind of threefold analysis suggested above. In particular, this confirms Leff's insistence on the need for interdisciplinary collaboration between a revised historical materialism and ecology, as well as other natural sciences. In particular, this form of analysis raises problems for any account, such as James O'Connor's, that seeks to identify a *contradiction* between capitalist forms and relations of production and their conditions. This approach also suggests that the question of capitalism's long-term ecological sustainability cannot be answered a priori from an economic theory of capital, but can only be addressed in terms of the relationship between the requirements of expanded capital accumulation and the possibilities for social and technical reorganization of labor processes as forms of appropriation of nature.

NOTE

1. See, e.g., Murray Bookchin, *The Ecology of Freedom* (Montreal: Black Rose Books, 1991).

Chapter 5

SOVIET ENVIRONMENTALISM
The Path Not Taken

ARRAN GARE

INTRODUCTION

Capitalism is a system that by its very nature must expand until it destroys the conditions of its own existence. It is hardly surprising, then, that Marxists in the Soviet Union argued that in the current environmental crisis lay the ultimate reason for replacing capitalism with socialism. As A. D. Ursal, the editor of *Philosophy and the Ecological Problems of Civilization*, argued:

> The crisis of the environment, which is reaching extreme development almost everywhere, coincides with the last stage of the general crisis of capitalism. A conviction is growing throughout the world that only collapse of the capitalist system and victory of socialism throughout the world will create a general, fundamental, social opportunity for rational use of natural resources and the highest degree of optimum interaction with nature. . . . Convincing evidence that socialism is a necessary condition for optimizing relations between society and nature is socialism as it actually exists, and the policy of socialist countries in respect of the environment.[1]

With the collapse of the Soviet Union, however, all hope that Soviet communism might be transformed into a more attractive, less environmentally degraded social order than the liberal democratic societies of the West has been destroyed. The description of the modern predicament by Alvin W. Gouldner has become even more poignant: "The political uniqueness of our own era then is this; we have lived and still live through a desperate political and social malaise, while at the same time we have *outlived* the desperate revolutionary remedies that had once been thought to solve them."[2] If this is the case, there is reason to examine the environmental failures of the Soviet Union more closely. Was it possible that things might have worked out differently? If so, does this provide any orientation for the present? In this chapter, I will show how an alternative path for Soviet society had been charted and partly implemented in the 1920s by the radical wing of Bolshevism, a path that made environmental conservation a central issue. And I will suggest that this is the path that holds most hope for the future.[3]

SOCIALIST ENVIRONMENTALISM

One of the unfortunate legacies of Soviet communism was to leave Russians ignorant of much of their own past. In the last decades of the Soviet Union, there emerged a large environmental movement.[4] This was more than a movement concerned with the environment. While some Soviet ideologists such as Ursal attempted to use environmental destruction in the West as an instrument of ideological struggle, and others such as Boris Komarov used this destruction to condemn communism as an inherently environmentally destructive system,[5] some saw in the environmental crisis a common cause for all humanity. Environmental destruction throughout the world was seen by Ivan Frolov (who under Gorbachev became editor of the Communist Party's theoretical journal *Kommunist*) to provide justification for ending the cold war, for reorganizing societies for the benefit of their members rather than for the struggle for world supremacy, and more fundamentally, for replacing anthropocentrism with "biocentrism" or "biosphere-ocentricism."[6] Since the overthrow of communism, new environmental movements have formed, mostly anti-Marxist either of a right-wing, extreme nationalist, and racist stripe, or of a left-wing, anarchist variety. However, none of these environmentalists appear to be aware that a strong environmental movement developed in the 1920s as one of the outcomes of the Revolution of 1917,[7] nor of the roots of this environmentalism in the ideas of the left wing of Bolshevism, a movement that attempted to create a synthesis of socialism and anarcho-syndicalism and which was aligned with Western Marxists opposed both to the control of society by markets and to the domination of society by centralized state bureaucracies.

The origins of environmentalism in Russia go back long before the revolution; also, there were a number of strands to Bolshevik environmentalism. In his monumental history of Soviet environmentalism up until 1935, Douglas Weiner pointed out the strong commitment by Lenin to the cause of conservation. In 1919, with Kochak's armies crossing the Urals and making their way toward the heartland of Soviet-controlled Russia, Lenin personally took time out from dealing with this crisis to hear the case for conservation.[8] Lenin's conservation policies and general attitude to government were for the most part very similar to those of the Progressive conservation movement that developed in the United States under Theodore Roosevelt.[9] Like Roosevelt, Lenin had a strong faith in science and was committed to creating an efficiently managed society. Lenin's environmentalism, while important and enlightened, offers us little that is new. In fact, there are good grounds for accepting the argument of the communist but anti–Bolshevik Anton Pannekoek that Leninism was simply the expression of the late drive by Russians for industrialization.[10] Marxism, as it was used by such Russians as Struve, Plekhanov, and Lenin, provided an ideology that enabled them (as it has since enabled a number of political leaders in the Third World) to appropriate the Western drive for technological development while struggling against efforts by the advanced capitalist societies of the West to subjugate them. The history of the Soviet Union has been a continuation of this struggle. It is impossible to understand the oppressive, technologically oriented policies of the Soviet Union except in relation to almost-constant threats of invasion from the West. However, many more radical ideas than those supported by Lenin were not only promoted, but also to some extent put into practice. The central theme of these radical ideas was that to create a socialist society it would be necessary to develop a new culture, which, among other things, would transform humanity's relationship to its environment.

In September 1918, the Proletarian Cultural and Educational Organizations, or Proletkul't, held its first All-Russian Conference to give substance to the dreams of these radical Marxists to create a proletarian culture. The leader of the radical Bolsheviks was Aleksander Aleksandrovich Bogdanov. To fully understand his ideas and their significance, it is necessary to see his work in relation to his political views and those of the philosophers and scientists whom Lenin and his fellow Bolsheviks condemned as idealists. These thinkers were influenced primarily by thermodynamics or energetics. Their "empiricism" was elaborated as part of their efforts to overcome the dualism between matter and mind associated with the mechanistic view of the world.[11] Being almost all socialists of one kind or another, they were among the founders of what Juan Martinez-

Alier has called "ecological economics." The first to develop ideas along these lines was a Ukrainian Narodnik, strongly influenced by Marx's economics, Sergei Podolinsky (1850–1891), who met Marx and Engels in 1872 and corresponded with Marx in 1880. Podolinsky tried to reformulate Marx's theory of surplus value in physicalist terms as appropriation of usable energy, thereby focusing attention on the limits of the natural environment, the way in which peasants were being exploited, and how some regions were being exploited by other ones. Such ideas were later reformulated, largely independently, by Edward Sacher, Leopold Pfaundler, Josef Popper-Lynkeus, Wilhelm Ostwald, Ernst Mach, Frederick Soddy, and Otto Neurath.[12]

BOGDANOV

Who, then, was Bogdanov? As a medical student, Bogdanov had become a Narodnik and still adhered to the views of Narodnaya Volya (the People's Will—the group that assassinated Alexander II in 1881) after his exile to Tula, his native town, in 1894.[13] It was while participating in political agitation in Tula that he became a Marxist. However, unlike most other Russian Marxists, Bogdanov was not interested in combating the Narodniks, and was sympathetic to the spontaneous action of the workers. In 1904, he wrote that "workers know better by experience what exploitation is" and urged the formation of trade unions and the use of strikes so that "the workers will unite in larger and larger masses."[14] After the uprising of 1905 in which workers with little direction from political leaders had almost succeeded in seizing power, he, along with a number of other Bolsheviks— including Maxim Gorky and Anatoly Lunacharsky—was strongly influenced by the ideas and practices of the anarcho-syndicalists.[15] He was particularly influenced by Georges Sorel (whose book *Reflections on Violence* was translated into Russian in 1907), who argued that what workers needed was a myth to inspire them to action rather than a scientific analysis of society. Bogdanov sympathized with Lunacharsky's efforts to join socialism with anarcho-syndicalism and his call for the subordination of political organizations to a class syndicalist organization, a kind of "General Worker's Soviet." This led in 1908 to the split in the Bolsheviks between Lenin and Bogdanov and his supporters, including Lunacharsky. It was this split that precipitated Lenin's attack on the philosophy of his opponents among the Bolsheviks in *Materialism and Empirio-criticism.* In 1909, Bogdanov wrote that "the working class as a social system does not exist unless the proletariat is organized into a party, syndicates, and so forth," as a "living collective."[16] Along with left Marxists of Western Europe such as Pannekoek and Gorter, Bogdanov extolled the work of the worker-philosopher Joseph Dietzgen (1826–1888), who had argued that "for a worker who seeks to take part in the self-emancipation of his class . . . the prime necessity is to cease allowing himself to be taught by others and to teach himself instead."[17] Dietzgen argued for a monist philosophy in which active, experiencing subjects had a place in the world. However, Bogdanov regarded Dietzgen's philosophy as still too much based on contemplation, defending Marx's (and modern physicists') concept of matter as that which resists labor (or action) against Dietzgen's conception of matter as primary being. The basic source of inspiration for Bogdanov's own philosophy, through which he was able to unite these diverse concerns, was the monism and energism of Wilhem Ostwald.

There were two main stages in Bogdanov's intellectual career. To begin with, in his work *Empiriomonism*, the second edition of which was published between 1904 and 1906, Bogdanov added a social dimension to the epistemological theories of the empirio-critics Ernst Mach and Richard Avenarius, whom Ostwald had used to justify taking energy, rather than matter, as the basic principle of scientific explanation. Opposing the empirio-critics' passive concept of experience, Bogdanov argued that the mental world is the product of individually organized experience, while the physical world is the product of socially organized experience. These two worlds reveal two different biological–organizational tendencies.[18] Like Western Marxists (and unlike Lenin), Bogdanov was interested in people's alienation from the world and from each other, and in the cultural conditions for creating a socialist society.[19] He argued that the conflicts of value associated with the sphere of individually organized experience were manifestations of the divisions within society based on class, race, sex, language, nationality, work specialization, and relations of domination and subordination of all kinds. It was necessary to overcome these conflicts to enable a new communal consciousness to emerge in which basic values could be agreed upon. But while Bogdanov accepted the idea that it was important to transform class relations to achieve this goal, he argued that this idea had been overemphasized by Marx. Other conflicts, including organizational relations and unequal relations between the sexes, also had to be overcome. To achieve this, the proletariat needed to transcend bourgeois culture, which he argued could only be achieved by creating a new culture to organize experience.[20] Bogdanov extended his critique of bourgeois culture to science. Anticipating later Marxist critiques of the science that emerged with capitalism, he saw the mechanical view of the world, the split between mind and matter, idealism and materialism, as expressions of the social practices of capitalist society, of the fetishism of commodities involved in market relationships, and of the division between the organizational and the executive functions in the labor process. Bogdanov called for a cultural regeneration based on developing the modes of understanding appropriate for a society in which the divisions in society, including the division between manual and mental labor, had been overcome.

The second stage of Bogdanov's intellectual career was devoted to providing the key to these modes of understanding. This was presented by him in his three-volume work, *Tektology: The Universal Oganizational Science*,[21] published between 1913 and 1922, in which Bogdanov developed the ideas of the energeticists in a second direction—as a general theory of organization. This new proletarian science was a precursor to, and possibly a superior version of, the systems theory of Ludwig von Bertalanffy.[22] Tektology was designed to provide a harmonious unity between the spiritual, the cultural, and the physical experiences of the "working collective," in whose interest all science and activity were to be organized and for whom all past culture, including bourgeois science, were to be reworked. By uniting the most disparate phenomena under one conceptual scheme, tektology would allow human beings torn apart by strife to find a common language. Since the sources of strife were larger than the merely economic, the common language had to be larger than traditional Marxism, although Marxism would be included as a special case. According to this philosophy, all objects are distinguishable as different degrees of organization. The focus was not on what the world was made of, but on the nature of organization. Organized complexes or systems are composed of interrelated elements, conceived of as activities, such that the whole is greater than the sum of its parts. Living beings and automatic machines are dynamically structured complexes in which "bi-regulators" provide for the maintenance of order. Bogdanov argued that no matter how different the various elements of the universe—whether electrons, atoms, things, people, ideas, planets, or stars—and regardless of the considerable differences in their combinations, it is possible to establish a small number of general methods by which any of these elements joins with another.

By conceiving humans as part of and within nature, as existing only through their capacity to obtain and process usable energy, Bogdanov brought the limitations of the natural environment into sharp focus. This concern was expressed in two novels written by Bogdanov to proselytize his ideas: *Red Star* and *Engineer Menni*.[23]

Both of these works were set on Mars, a planet where the communist order had already been established and society was governed by a "Council of Syndicates." *Red Star*, written in 1908, is the story of Leonid, a communist revolutionary who after the attempted revolution of 1905 is taken to Mars. There he is at first impressed by the harmony and fullness of life brought about by the communist revolution. It soon appears, however, that this harmony is superficial, and that Mars is suffering from the effects of its successes. Industries have become so dangerously polluting that many have to be relocated underground. The population is growing so rapidly that food shortages and even famines are predicted within several decades. Natural resources, including the radioactive matter that is the main source of energy, are being exhausted. Forests are being destroyed. Most importantly, socialist Mars has created a "Colonial Group" in its government and is preparing to create colonies on Earth or Venus to replenish Mars's resources. The highlight of *Red Star* is a debate between two Martians over whether they should exterminate the Earthlings to get access to more natural resources. Sterni takes the position that Earthlings are so hopelessly malformed by their evolutionary past that even the Earth's socialist minority would never be able to work together amicably with their fellow socialists on Mars. To prevent a long guerrilla war of resistance, Sterni argues that the Earthlings should be wiped out in advance, painlessly, via death rays, so that the riches of Earth can be used to build a more humane socialism on Mars. Netti, Sterni's opponent, reprimands him for proposing to eliminate "an entire individual type of life, a type which we can never resurrect or replace."[24] Sterni, according to Netti, "would drain forever this stormy but beautiful ocean of life." He does not recognize that "the Earthlings are not the same as we. They and their civilization are not simply lower and weaker than ours—they are *different*."[25]

THE PROLETKUL'T MOVEMENT

The Proletkul't movement, inspired by Bogdanov and largely under his direction, at its height boasted 400,000 members, was publishing twenty journals, and had attracted the support of a wide spectrum of Russia's artists, musicians and writers.[26] In 1919, Bogdanov also established a proletarian university in Moscow with four hundred students. People in part inspired by Proletkul't formed the "Worker's Opposition," which opposed the bureaucratic tendencies of the new government, the "return to capitalism" of the New Economic Policy (N.E.P.), and Trotsky's call to militarize society. Instead, the Worker's Opposition called for worker control of industry. This whole development, which resonated with developments in Western Marxism, was attacked by Lenin, who saw it as a syndicalist threat to his own political philosophy and the institutions he was building.[27] Lenin, who conceived history in dualist terms as a dialectical conflict between spontaneity and conscious direction, in which progress is achieved through the control of spontaneity by consciousness,[28] condemned the syndicalist tendencies among Marxists as an "infantile disorder."[29] Bogdanov in particular came under attack. Lenin, as Robert Williams has noted, "was well aware that behind the Aesopian language of 'experience,' 'energy,' and 'collectivism' lay the syndicalist politics of direct

action,"[30] and republished his *Materialism and Empirio-criticism* to undermine Bogdanov's authority. In line with his philosophy of history and
political philosophy, this work affirmed a fundamental dualism between
consciousness and the world, with knowledge being conceived as the true
representation of the world. Late in 1920, Lenin forced the subordination
of the hitherto freewheeling Proletkul't to the People's Commissariat of
Enlightenment or Education (Narkompros), and the former was soon
abolished altogether. By the time *Tektology* was completed in 1922, Bogdanov's prestige had been all-but destroyed. However, his works continued
to have an influence, particularly through the commissar of enlightenment,
Anatoly Vasilyevich Lunacharsky, Bogdanov's brother-in-law and a sup-
porter of his philosophy.[31]

Lunacharsky had become the commissar of enlightenment in 1917
and remained in this position until he resigned in September 1929. This
period is regarded as the Golden Age of Soviet culture, largely due to the
influence of the Commissariat for Enlightenment and the policies promoted by Lunacharsky. These achievements can be accounted for by the
increased state support for education and other cultural activities, by the
pluralistic policies pursued by Lunacharsky, and also by the significance
accorded to culture, and correspondingly, to the intense debates on
fundamental issues of culture. These debates progressively impinged upon
the sciences.

SOVIET SCIENCE

Alexander Vucinich in his study of the Academy of Sciences of the Soviet
Union characterized scientists of the 1920s as struggling to rebuild science
after the chaos of the Great War and the Civil War, and to fend off Marxist
efforts to control science. He claimed that "during the first ten years under
Soviet rule the Academy was involved in a gruelling struggle to regain the
growth momentum lost at the beginning of World War I: it was not until
1928 that its publication output reached the prewar level."[32] For Vucinich,
it was only with the Stalinization of Soviet science, the reduction of science
to an instrument of the economy, that science came to be Marxist. I wish
to suggest that it was the developments in science that took place under
the auspices of Lunacharsky's Commissariat of Enlightenment that give us
some idea of what a true socialist science would be like. Later developments
are better characterized as revivals of Russian nihilism.[33] Developments in
science in the 1920s were moving Soviet society toward a new relationship
to its natural environment, and these developments were closely associated
with the conservation movement.

Initially, the Commissariat of Enlightenment promoted the establishment of specialized Institutes of Research, cultivating the support of the largely anti-Marxist scientific establishment. Marxist appointees within universities defended science as the ultimate product of human consciousness. Their views of science were essentially positivistic; science was seen as superior to and independent of philosophy, and as mechanistic. Reductionist theories, such as Pavlov's psychology, were defended, and it was argued by these Marxists (and by Trotsky) that the goal of science is to explain the world in terms of chemistry and physics. However, these "mechanists" were soon opposed by a new intellectual movement. In 1918, the communist government set up the Socialist Academy of the Social Sciences, renamed in 1923 as the Communist Academy, which rapidly expanded its activities, becoming the guiding star in wide-ranging efforts to create new centers for the training of future Marxist scholars. In 1921, the Academy set up the Institute of the Red Professoriat to supply institutes of higher learning with Marxist instructors in economics, sociology, and philosophy, thereby providing the conditions for the establishment of a Marxist intellectual culture. In 1924, the Society of Militant Dialectical Materialists was founded; its leader, A. M. Deborin, based in the Communist Academy and giving seminars at the Institute of the Red Professoriat, was able to create a movement devoted to critically scrutinizing the philosophical assumptions of natural science.[34] Bogdanov's idea of a proletarian science was refurbished.

While the mechanists claimed that the successes of reductionist science validated their position, the dialecticians were strengthened by the publication in 1925 of Engels's *Dialectics of Nature*. The dialecticians rejected both the reductionism of the mechanists and the organic analogies of Western antimechanists. While rejecting Bogdanov's philosophy as such, the dialecticians nonetheless argued that nature is essentially dynamic and creative, generating qualitatively new processes that cannot be understood in terms of the conditions of their emergence. Humans were seen as creative participants within nature, able to control their own destinies. From 1926 onward, the dialecticians not only criticized developments within science, but were able to influence the direction of scientific research.

Following Stalin's alliance with the supporters of the N.E.P. in 1928, one aim of which was to expel Trotsky, Kamenev, and Zinoviev from the party leadership, Stalin embraced the cause of workers who, disaffected by the contrast between the decline of their own living standards and the growing prosperity of the peasantry, regarded the N.E.P. as a betrayal of the revolution. Responding to the demands of these workers, he initiated a cultural revolution to purge society of bourgeois forms of thought.[35] Initially, this move put the Deborinites in a good position to exert their influence, which culminated in 1929 when they gained control of the

Communist Academy and other institutions. Entire fields of science were then scrutinized for their philosophical assumptions.

Stalin's main agenda, however, was to speed up economic growth. Arguing that immediate industrialization was required to face the growing threat from Western Europe and that collectivization of the land was required to supply the growing number of workers with food, Stalin called for a reassertion of conscious direction over spontaneity. Stalin was working to destroy the power of the remaining Bolshevik leaders[36] and Deborin's ideas lost their support. Having breached the walls of the "bourgeois professoriat," and having established the principle of direct political intervention in scientific institutions, Deborin and his followers were attacked by former Deborinites, led by M. B. Mitin, for not serving the revolution. By the end of 1930, by which time Lunacharsky had resigned as commissar of enlightenment in protest against the rejection of his ideals of humanistic education and cultural pluralism, the Deborinites were out of favor.[37] While "mechanists" had been knowledgeable about science, but ignorant of philosophy, and the Deborinites had been knowledgeable about philosophy, but ignorant of science, Mitin and his followers achieved a dialectical synthesis of the ignorance of each. However, there was more to Mitin's views than ignorance. He revived the ideas of the nihilists of the 1860s and 1870s, in particular, the idea that science is nothing but an instrument for the development of technology. It was Mitin's defense of this view that endeared him to Stalin, who then dismissed the Deborinites as "Menshevizing idealists," his ultimate term of abuse and dismissal. Thereafter proletarian science was no longer antimechanistic science, but science in the service of the Five-Year Plans devoted to the domination of nature.

THE CAREER OF ECOLOGY

Prior to the revolution there had been a range of environmentalists in Russia roughly corresponding to the range found in Western Europe and the United States.[38] First, there were those who were concerned about environmental destruction for purely utilitarian reasons, evaluating nature only as an economic resource. Second, there were those who extolled the intrinsic value of nature, and called for a recognition of the rights of all living things to their existence, for example, Ivan Parfen'evich Borodin and Andrei Petrovich Semenov-tian-shanskii. There was also a third group, a scientific one based on the development of phytosociology, the study of vegetational communities. These pioneers of plant ecology looked to virgin nature as a model of harmony, efficiency, and productivity that agriculturalists should strive to emulate. It was argued that to put agriculture on a sound basis pristine natural communities should be studied; it was also proposed that areas of nature be set aside as models (*etaloni*) within protected nature reserves (*zapovedniki*) against which cultivated land could be compared. It was this third group that gained vigorous support after the revolution, first from Lunacharsky who commended the idea to Lenin, and then by Lenin who did all he could to support the environmentalists. Following Lenin's support for the first proposed *zapovednik* in 1919, responsibility for their creation and administration was granted by Lenin to Lunacharsky's Commissariat of Enlightenment to ensure independence from short-term economic imperatives. By 1929, sixty-one *zapovedniki* had been established, with a total area of almost four million hectares, distributed throughout the Soviet Union to provide the basis for developing a comprehensive understanding of the natural environment of the whole country.[39] After a number of battles with the Commissariat for Agriculture (Narkomzem) and the Commissariat for Trade (Narkomtorg), these *zapovedniki* were able to support a rapid expansion of ecology. Associated with this development, ecology was increasingly included in the curricula of universities and schools, and in the later 1920s ecologists were able to make a determined effort to influence state economic policy.[40]

Before the revolution, Russian ecology had focused almost exclusively on plants and soils. With the provision of *zapovedniki*, Soviet ecologists began to appreciate the role of fauna in shaping the development of natural communities, which were seen as complex systems of three interacting elements of equal importance: vegetation, fauna, and the abiotic environment. Within this framework, a diversity of theories were elaborated. By 1931, when Daniil Nikolaevich Kashkarov published his great survey textbook of community ecology, *Environment and Community* (later published in English), it could be fairly argued that the Soviet Union led the world in ecology. To appreciate some of the ideas being developed by Soviet ecologists, and how these developments were related to the communist revolution and to the ideas of Proletkul't, it is only necessary to look at the career of the man who had by 1931 become one of the foremost ecologists in the Soviet Union, Vladimir Vladimirovich Stanchinskii.

Stanchinskii obtained his doctorate from Heidelberg University in 1906, then found that it was not recognized in Russia, and hence had to pass external exams at Moscow University.[41] It was only after the revolution, in the new, intellectually freer environment created by the Commissariat of Enlightenment, that Stanchinskii was able to embark on a successful career. During the Civil War, he headed the local El'ninsk district branch of Narkompros (Commissariat of Enlightenment) of the Russian Socialist Federated Republic (RSFSR) in Smolensk Oblast, and was one of the organizers of the new Smolensk University set up by Narkompros. Playing a major role in Smolensk intellectual life, he became a full professor at

Smolensk University and head of its Department of Zoology, while also serving as the president of the Smolensk Society of Physicians and Naturalists, which he founded. Having an exceptionally broad vision, he soon gravitated to one of the leading theoretical problems in biology, the mechanism of speciation. He then moved on to what had been defined in the Soviet Union as the other great theoretical issue of the day: the nature of biological community. His guiding idea in this study, an idea clearly resonating with Bogdanov's energistic philosophy, was that by virtue of their being in a continual state of matter-and-energy exchange with their environment, and continually changing, destroying, and synthesizing substances within themselves, each species must be seen to have a very specific biochemical and physicochemical role in the "economy of nature." Stanchinskii had visited the *zapovednik* at Askania-Nova in 1926 and thereafter decided this was an ideal spot to relocate his investigations into biological communities. In the spring of 1929, he assumed the posts of deputy director of the reserve and director of its scientific sector, simultaneously gaining appointment as head of the Department of Vertebrate Zoology at Khar'kov University.

Biological communities had previously been defined by their floristic composition, by certain structural features, or by a certain visual homogeneity. Stanchinskii investigated food webs to identify the boundaries of communities within nature, tracing the transformation of solar energy by vegetation and other autotrophes through myriad biotic pathways until all the accumulated energy potential had been exhausted. He demonstrated how the biocenosis (biological community) was characterized by relative stability, a "dynamic equilibrium" in which relative numbers of the various component species remained surprisingly constant over long periods of time, despite their theoretical ability to propagate exponentially. Placing each organism on a "trophic ladder," Stanchinskii showed how each successive rung of the ladder would have less energy in the form of food than the next lower level, since each successive level was dependent on the previous one for its energy supply, and energy was dissipated at each level. He then constructed an ideal mathematical model to describe the annual energy budget of a simple theoretical biocenosis to guide his empirical research, developing a methodology and an instrumentation for measuring the biomass of the various component species of a biocenosis.

What is significant about Stanchinskii's career is not simply the ideas he developed, which are now recognized to have been about ten years in advance of the work of American ecologists (whom he influenced), but the way in which his career was made possible by changes wrought by the communist government, and the status his ideas were accorded within the Soviet Union. It appears unlikely that Stanchinskii's career would have been possible without the new opportunities opened by the expansion of education, and the establishment of new scientific institutions and of the *zapovedniki* inaugurated by the communist government. It also appears unlikely that the diversity of theoretical approaches to ecology developed in the 1920s and early 1930s could have taken place in the rigid institutions of prewar Russia. The cultural flowering of the 1920s, of which the development of ecology was a part, can only be accounted for by the ferment created by the significance accorded to culture, particularly to science, and the Marxist challenges to the assumptions underlying the sciences. This seems particularly evident in the case of Stanchinskii's work. Stanchinskii was inspired by Vladimir Ivanovich Vernadskii,[42] a prominent member of the Academy of Science who was unsympathetic to Marxism, although sympathetic to the work of Podolinsky. Vernadskii's work on geochemistry and biogeology, which led him to promote and elaborate the concept of the biosphere, were entirely in accordance with Bogdanov's tektology, and Proletkul't and the Commissariat of Enlightenment had created a sympathetic environment for such ideas. So while Vernadskii was criticized by some Marxists, his concepts were accepted into the mainstream of science in the Soviet Union in a way that contrasts radically with the marginal place similar ideas have occupied in the West up to the present.[43] The favorable reception of Stanchinskii's ideas can also be accounted for by the intellectual environment created by Bogdanov's philosophy and the Proletkul't movement. The high status accorded to science, and to ecology within science, particularly when ecology was formulated in terms of energetics, gave Stanchinskii a significance in Soviet society unmatched by ecologists in other countries.

This high status attracted the attention of the Deborinites. I. I. Bugaev of the Communist Academy, who was assigned the task of investigating the ecologists, attacked those ideas that failed to allow for emergence, and thereby the irreducibility of humanity to biology. Pachoskii's attempt to prove the necessity of inequality in nature and to extend this to humanity was censured. But Stanchinskii was able to reformulate his ideas to accord with the strictures of the Deborinites, and arguably strengthened his theory and his research program in the process. He replaced the static-sounding notion of the equilibrium of the biocenosis with the notion of "proportionality" and emphasized the continuous self-creation of the biocenosis, which he depicted as growing out of interactions among its components and between them and the abiotic environment, with the result that new syntheses were continually arising in the form of successional series.[44] His work was then not only acceptable to the Deborinites, but could be taken by them as further corroboration of the dialectical nature of the world.

With the backing of his ecological theory, Stanchinskii was able to argue for a role for ecology in economics. He argued that by studying the energy flows in a whole range of biocenoses, humans would be able to

calculate the productive capacities of these natural communities and would
be able to structure their own economic activity in conformity with them.
He also saw such a program of biocenotic research as an aid in achieving
biotic protection of cultivated croplands, thereby overcoming the need to
use harmful pesticides. Stanchinskii played a major part at the First
All-Russian Congress for the Conservation of Nature, held in September
1929, where he argued that ecologists must play a major part in the
formulation of the Five-Year Plan, arguing that conservation organizations
must be able to review plan targets and monitor plan fulfillment. The
congress accepted his arguments and resolved:

> The economic activity of man is always one form or another of the
> exploitation of natural resources. . . . The distinction and tempo of
> economic growth can be correctly determined *only* after the detailed
> study of the environment and the evaluation of its production capacities
> with the aim of its conservation, development and enrichment. This is
> what conservation is all about.[45]

However, with Stalin's intervention into the cultural sphere, in support
of his drive for industrialization, individuals and organizations not whole-
heartedly behind this drive were placed in a precarious position. The whole
science of ecology was affected by this. The ecologists failed in their effort
to gain a place in economic planning within the Soviet Union. They
nevertheless became the most trenchant critics of the implementation of
the Five-Year Plan. They opposed the damming of rivers without due care
for the ecological effects; the collectivization and uniform mechanization
of agriculture; the efforts to acclimatize exotic fauna; and interference in
the life-styles of traditional societies occupying ecologically fragile environments. But, they, in turn, drew a massive response from the Stalinists, who
condemned the conservationists as "organically alien to active youth and
in particular to Soviet Youth, seized . . . with the enthusiasm of socialist
construction and reconstruction."[46] V. L. Komarov argued in 1931 that all
reference to "plant communities" should be expunged from biology, a call
which foreshadowed a drive against ecology by I. I. Prezent, a colleague of
Lysenko, who was committed to the wholesale acclimatization of exotic
species and creating a world in which "all living nature will live, thrive, and
die at none other than the will of man and according to his designs."[47]
Stanchinskii lost his job, the research station at Askania-Nova was closed
down, and in 1934 he was arrested. The typesetting for his book which was
about to be published was destroyed. Although conservationists fought a
rearguard action, this was eventually defeated, and the science of ecology
was virtually suspended for two decades.

CONCLUSION

The story of Proletkul't, Bogdanov, Lunacharsky, and the ecologists is the story of the path not taken. But it was a path sufficiently ventured upon to show what might have been if Leninism, followed by Stalinism, had not triumphed. It is this untaken path, the path of cultural revolution on the basis of a postdualist (and postmechanist) conception of the world in which people are seen as active, conscious participants within nature rather than as standing over and above nature, conjoined with a struggle to transform the social order that engendered such dualism—the commodification of the world, the division between intellectual and manual labor, and relationships between people based on domination and subordination—which modern environmentalists must now consider.

NOTES

1. A. D. Ursal, ed., *Philosophy and the Ecological Problems of Civilisation*, trans. H. Cambell Creighton (Moscow: Progress Publishers, 1983), 10fn.

2. Cited without reference in Alec Nove, *The Economics of Feasible Socialism* (London: George Allen & Unwin, 1983).

3. This is further argued in Arran E. Gare, *Beyond European Civilization: Marxism, Process Philosophy, and the Environment* (Bungendore, Australia: Eco-logical Press, 1993).

4. See Philip R. Pryde, *Conservation in the Soviet Union* (Cambridge: Cambridge University Press, 1972), and *Environmental Management in the Soviet Union* (Cambridge: Cambridge University Press, 1991); Joan DeBardeleben, *The Environment and Marxism-Leninism: The Soviet and East German Experience* (Boulder, Colo.: Westview Press, 1985); and Douglas R. Weiner, "The Changing Face of Soviet Conservation," in *The Ends of the Earth: Perspectives on Modern Environmental History*, ed. Donald Worster (Cambridge: Cambridge University Press, 1988), 252–273.

5. Boris Komarov, *The Destruction of Nature in the Soviet Union* (London: Pluto Press, 1978). The author now lives in Israel.

6. Ivan T. Frolov, "The Marxist-Leninist Conception of the Ecological Problem," in Ursal, ed., *Philosophy;* and I. T. Frolov and V. A. Los, "Filosofskie osnovaniia sovremenio ekologii," *Ekologischeskaia propaganda v SSSR* (Moscow: Nauka, 1984). This paper is discussed and partly translated by Douglas Weiner in "Prometheus Rechained: Ecology and Conservation in the Soviet Union," in *Humanistic Dimensions of Science and Technology in the Soviet Union*, ed. Loren R. Graham (Cambridge, Mass.: Harvard University Press, 1989).

7. Douglas Weiner gave a paper he had written on Stanchinskii to Frolov in 1985 when Frolov spoke at Boston University. Frolov appeared to be unaware of the conservationists of the 1920s.

8. Douglas R. Weiner, *Models of Nature: Ecology, Conservation, and Cultural Revolution in Soviet Russia* (Bloomington: Indiana University Press, 1988), 27.

9. This has been described by Samuel Hays in his *Conservation and the Gospel of Efficiency: The Progressive Conservation Movement, 1890–1920* (Cambridge, Mass.: Harvard University Press, 1959).

10. Anton Pannekoek, *Lenin as Philosopher* (1932; reprint, London: Merlin Press, 1975).

11. Seen in relation to the history of logical positivism, their philosophies have been largely misrepresented. Paul Feyerabend on reading Mach has characterized him as a "dialectical rationalist"; see "Mach's Theory of Research and Its Relation to Einstein," in *Farewell to Reason* (London: Verso, 1987), 192–218.

12. J. Martinez-Alier and J. M. Naredo, "A Marxist Precursor of Energy Economics: Podolinsky," *Journal of Peasant Studies, 9,* January, 1982; and Juan Martinez-Alier, *Ecological Economics* (Oxford: Basil Blackwell, 1987).

13. On the early evolution of Bogdanov's political ideas, see James D. White, "Bogdanov in Tula," *Studies in Soviet Thought, 22,* February, 1981. See also John Biggart, "Bogdanov and Lunacharskii in Vologda," *Sbornik, 5,* 1980.

14. Cited by Robert C. Williams, in "Collective Immortality: The Syndicalist Origins of Proletarian Culture," *Slavic Review, 39,* September, 1980, 395, from Riadovoi [Bogdanov], *O sotsializme* (Geneva: 1904), 15, 17, 21.

15. On the syndicalist influence on Bogdanov and other Marxists, see Williams, "Collective Immortality," 389–402. However, none of the Marxists appear to have been influenced by the work of Peter Kropotkin.

16. Cited from A. Bogdanov, "Filosofiia sovremennago estestvo ispytatelia," in *Ocherki filosofi kollektivizma* (St. Petersberg, 1909), 133, by White, "Bogdanov," 397.

17. Cited by D. A. Smart, *Pannekoek and Gorter's Marxism* (London: Pluto Press, 1978), 18.

18. See Kenneth Jensen, *Beyond Marx and Mach: Aleksander Bogdanov's Philosophy of Living Experience* (Dordrecht, Holland: D. Reidel, 1978).

19. See Zenovia A. Sochor, *Revolution and Culture: The Bogdanov–Lenin Controversy* (Ithaca, N.Y.: Cornell University, 1988).

20. It appears that Gramsci's ideas ultimately derived from Bogdanov. See Zenovia A. Sochor, "Was Bogdanov Russia's Answer to Gramsci?," *Studies in Soviet Thought, 22,* February, 1981.

21. This has not been translated. However a good idea of his philosophy can be gained from *Essays in Tektology: The General Science of Organization,* trans. George Gorelik (Seaside, Calif.: Intersystems Publications, 1980). See also George Gorelik, "Bogdanov's Tektology: Its Basic Concepts and Relevance to Modern Generalizing Sciences," *Human Systems Management, 1,* December, 1980.

22. On the relationship between tektology and systems theory, see Ilmari Susiluoto, *The Origins and Development of Systems Thinking in the Soviet Union: Political and Philosophical Controversies from Bogdanov and Bukharin to Present-Day Re-evaluations* (Helsinki: Suomalainen Tiedeakatemia, 1982); and George Gorelik, "Bogdanov's Tektology: Its Nature, Development, and Influence," *Studies in Soviet Thought, 26,* July, 1983.

23. Alexander Bogdanov, *Red Star: The First Bolshevik Utopia*, ed. Loren R. Graham and Richard Stites, trans. Charles Rougle (Bloomington: Indiana University Press, 1984).

24. Ibid., 116.

25. Ibid., 117.

26. See Lynn Malley, *Culture of the Future: On the Proletkult Movement in Revolutionary Russia* (Berkeley and Los Angeles: University of California Press, 1990).

27. Peter Kropotkin had written to Lenin, predicting that "the syndicalist movement . . . will emerge as the great force over the next fifty years, leading to the creation of the communist stateless society" (cited by Robert C. Williams, "Childhood Diseases: Lenin on 'Left' Bolshevism," *Sbornik, 8,* 1982, without a reference). In 1920, there had been an epidemic of trade union unrest, anarchists had bombed the Moscow party headquarters, and in the Ukraine the anarchist Makhnovites, flying the black flag, were attacking the Red armies as well as the White. Toward the end of 1920, anarchists organizations in Russia were crushed.

28. The best analysis of Lenin's political philosophy is Neil Harding's, *Lenin's Political Thought* (Atlantic Highlands: Humanities Press, 1983). However it is in Katarina Clark's, *The Soviet Novel: History as Ritual* (Chicago: University of Chicago Press, 1981) that the best analysis of the dialectic of spontaneity and consciousness and the influence of this on Soviet culture are provided.

29. V. I. Lenin, " 'Left-Wing' Communism—An Infantile Disorder" (April 1920), in Robert C. Tucker, ed., *The Lenin Anthology* (New York: Norton, 1975). On Lenin's relation to left Marxists, see Williams, "Childhood Diseases."

30. Williams, "Collective Immortality," 397.

31. On Lunacharsky's policies and influence, see Sheila Fitzpatrick, *The Commissariat of Enlightenment* (Cambridge: Cambridge University Press, 1970).

32. Alexander Vucinich, *Empire of Reason* (Berkeley and Los Angeles: University of California Press, 1984), 122.

33. That Stalinism was a conscious throwback to the nihilism of the 1860s has been argued by James H. Billington *(The Icon and the Axe* [New York: Vintage Books, 1970; originally published in 1966], 534fn.). The connection between Stalinist biology and the biologists of the 1840s–1860s (extolled by the Nihilists) has been pointed out by Douglas Weiner ("The Roots of 'Michurinism': Transformist Biology and Academicitization as Currents in Russian Life Sciences," *Annals of Science, 42,* 1985).

34. The conflict between the mechanists and the Deborinites has been described by David Joravsky, in *Soviet Marxism and Natural Science 1917–1932* (New York: Columbia University Press, 1961).

35. See Sheila Fitzpatrick, "Cultural Revolution as Class War" and Moshe Lewin, "Society, State, and Ideology during the First Five-Year Plan," in *Cultural Revolution in Russia, 1928–1931,* ed. Sheila Fitzpatrick (Bloomington: Indiana University Press, 1978).

36. Stalin's main opponent was Bukharin, whose ideas were influenced by Bogdanov's work. Had Bukharin survived and defeated Stalin, the position of the conservationists would have been much more secure. On Bukharin, his relation

to Bogdanov, his ideas and political career, and his struggle with Stalin, see Stephen F. Cohen, *Bukharin and the Bolshevik Revolution: A Political Biography, 1888–1938* (Oxford: Oxford University Press, 1980).

37. See Leszek Kolakowski, *Main Currents of Marxism, Vol. 3: The Breakdown,* trans. P. S. Falla (Oxford: Oxford University Press, 1981).

38. On the early history of Soviet ecology and conservation, see Weiner, *Models of Nature.* I am deeply indebted to Professor Weiner for sharing with me his vast knowledge of Russian ecology and environmentalism, and indeed for his help in my efforts to understand the dynamics of Russian culture.

39. Ibid., 61.

40. On the rapid expansion of ecology in Russia in the 1920s, and the rapidity with which ecology entered the curricula of universities, see (apart from the work of Weiner) J. Richard Carpenter, "Special Review: Recent Russian Work on Community Ecology," *Journal of Animal Ecology, 8,* 1939.

41. For details on the life and career of Stanchinskii, see Weiner, 1988, *Models of Nature,* 78–82 and passim.

42. On Stanchinskii's relation to Vernadskii, see ibid., 80.

43. See Lynn Margulis and Edward Goldsmith, "Discussion," in *Gaia: The Thesis, the Mechanisms, and the Implications* ed. Peter Bunyard and Edward Goldsmith (Camelford, U.K.: Wadebridge Ecological Society Centre, 1988), 166.

44. Douglas Weiner, "Community Ecology in Stalin's Russia," *Isis, 75,* 1984, 684–696, 692.

45. Douglas Weiner, *The History of the Conservation Movement in Russia and the U.S.S.R. from its Origin to the Stalin Period* (Ph.D. thesis, Columbia University, 1983), 348.

46. Ibid., 275.

47. Ibid., 517.

Chapter 6

MARXISM AND ECOLOGY

Gunnar Skirbekk

Only recently has ecology become an object of study and concern, and it has not yet made real inroads into politics, even if this winter's oil [1974–1975] crisis has begun to alert the public to ecology's significance. It must be said that the training of political leaders is more strongly grounded in philosophy and political science than in biology, and furthermore that the established interests hinder any serious political concern with ecology. The time has come to make politics and ecology interact. This is what I would like to attempt concerning Marxism, and I believe this effort will result in a mutual clarification of the two systems.

As a preliminary, ecology has to be situated in a long-term perspective. If it is only a matter of eliminating the pollution of the Seine within five years, the task is feasible; capitalism *is* capable of resolving limited ecological problems. But what really is at issue is a global and enduring challenge that neither capitalism nor Soviet socialism is prepared to face. I should point out that I do not believe in a planetary "ecocatastrophe," but rather in an "ecocrisis" that will affect various classes and countries in various forms at various times. If by "capitalism" we mean an economic system that seeks to increase profit in an open market, this system will become impossible as soon as, for ecological reasons, a future society demands political control of the economy. In the long-term perspective, the problem

129

is therefore not "capitalism or socialism," but "socialism or fascism," that is, global democratic and egalitarian control, or global totalitarian, authoritarian, and probably racist control. The Soviet regime does not supply us with the model for this indispensable socialism, since it also exploits resources with no concern for ecology.

If these considerations are correct, what are their implications for Marxism?

Marx touched upon the problem of ecology:

> Capitalist production collects the population together in great centers, and causes the urban population to achieve an ever-growing preponderance. This has two results. On the one hand it concentrates the historical motive power of society; on the other hand, it disturbs the metabolic interaction between man and the earth, i.e., it prevents the return to the soil of its constituent elements consumed by man in the form of food and clothing; hence it hinders the operation of the eternal natural condition for the lasting fertility of the soil. . . . All progress in increasing the fertility of the soil for a given time is a progress towards ruining the more long-lasting sources of that fertility. The more a country proceeds from large-scale industry as the background of its development, as in the case of the United States, the more rapid is this process of destruction. Capitalist production, therefore, only develops the techniques and the degree of combination of the social process of production by simultaneously undermining the original sources of all wealth the soil and the worker.[1]

We note, among other things, that Marx speaks of the *Naturbedingung*, the "natural condition."[*]

However, for historical reasons, ecology as a research discipline and ecology as a crisis were almost unknown to Marx, even if some of the problems that are today called "ecological" were part of the proletarian misery that Marx knew and described so well: malnutrition, polluted air and water, noise, the degenerating environment, overpopulation. What is new is the universal character of these problems, and the fact that today certain elements of the bourgeoisie should be distressed about it; what is new above all is that the direction and seriousness of current developments represent a global danger of which Marx, in his time, was ignorant, and

---

[*]Translator's note. In a footnote, the author notes the difficulty of translating *Bedingung* (and its plural, *Bedingungen*) into French, opting (following J. J. Lentz) for "*conditions existantes.*" In parentheses that follow "*conditions existantes*" in the body of the text, he remarks that this expression has been omitted from the Pléiade translation of *Capital*. For all of Skirbekk's subsequent references to this expression, I have maintained the singular (*Bedingung*) form that Marx uses in the passage cited above, and adopted Ben Fowkes's translation, "natural condition."

that his theory of the future development of capitalism did not allow him to foresee. For according to Marxian theory, capitalist exploitation of man and capitalist impoverishment of the earth would be overcome by a rational and international control that socialism would impose fairly quickly, since capitalism entails pauperization and results in crises of overproduction and revolution. For the moment, however, capitalism has more or less succeeded in resisting crises of overproduction by increasing the consumption of a large part of the masses in the advanced countries. Thus capitalist productivity has assumed dimensions unforeseen by Marx, while creating the current ecological crisis. Liberal capitalism, transformed into a planned capitalism, builds schools, hospitals, and social institutions. In collaboration with unions, capitalists pursue the development of production and consumption, which imposes a desire for rational planning of salaries and of state and private investments.

The growing ecocrisis of capitalist countries can apparently be analyzed in Marxian terms, but only in Marxian terms ecologically reframed: capitalism is moving toward a self-destructive crisis, not the one that was forecast, but toward an even more serious crisis, a universal ecocrisis. And this is why it is necessary to rethink certain elements of Marxism.

The conception of "natural condition" needs to be reevaluated. The *infrastructure* is comprised not only of the productive forces and the relationships of production, but also of the forces of nature. This implies that a reconciliation between the productive forces and the relationships of production, that is, a traditional socialist revolution, is not enough: there also must be a reconciliation between these two factors and the forces of nature; the socialist economy must be ecological.

For Marx the relationship of dialectical tension between the productive forces and the relationships of production constituted the motive force of history. The existing conditions of nature remained, for him, invariable conditions of production, that is, a constant, static element in social development. Natural resources are not known as limited, and they thus may form part of the static framework of production.

The existing conditions of nature will therefore have another position in the ecological perspective. Technoindustrial development has gradually created an opposition between the productive forces and the forces of nature, an opposition that will determine the future development of the world in a decisive fashion. Thus the opposition between the productive forces and the relationships of production can no longer be considered as the sole fundamental element of historical development.

The ecological perspective also plays a role in the theory of *surplus value*. Under the capitalist regime, value is created by productive workers. The price of commodities is composed of the cost of raw materials, of the reproduction of the forces of production, of administration and taxes, of the cost of reproduction of the workers and their families in the current social conditions, and finally of surplus value, the profit which, through the reinvestments that competition makes necessary, implies capitalist accumulation and expansion. In this sense, surplus value represents the nucleus of capitalism, the principle of its exploitation and its objective injustice. Let me stress that value is created by labor, and only by labor. Natural resources are certainly necessary for labor, but resources do not create value. Resources represent a source of wealth (wealth that is not simply a natural fact, but one that depends upon the technological and scientific development of society). So far, so good. Nevertheless, in an extractive capitalist economy not only are workers exploited, but resources are destroyed. If you will, nature is impoverished through the destruction of resources as resources; a part of the natural resources is used without being restored, that is, without an equal quantity of wealth being returned to nature. Since resources are limited and the world population will continue to grow, this destructive extraction of limited natural wealth represents an impoverishment of future generations.

In order to do something constructive, one can try to put a price on resources, which today cost too little. As we know, this is a delicate problem, theoretically and practically. How can the loss of a beautiful site or of an animal species be evaluated, in relationship to a certain production of electricity or fuel? Finding a common criterion for incommensurable phenomena is a theoretical dilemma. And what will be the result if we pay the complete ecoprice, which may mean the inability of some companies to withstand competition, and hence bankruptcies, dismissals, and an unemployment crisis. The market economy displays its limits.

Let us use our imagination, and attempt to evaluate an ecological price. We shall add to the ordinary market price the cost, also in terms of market price, of cleaning up the pollution produced by each commodity, by putting the recyclable elements of each commodity back into circulation, and of restoring the ecosystems damaged by each commodity. And in the price, by our taxes, can be added the expenditures caused by use and decomposition of each commodity. This is feasible in a way. But, continuing our supposition, an ecoprice may more or less reduce profitability. In a first stage, the profit from a commodity is reduced, but the production of the commodity remains profitable. The ecoprice eliminates that profit. Production is healthy, from an ecological point of view, but this production can no longer function capitalistically, unless the same elimination of profit is simultaneously introduced into the whole system, thus eliminating competition. Then the ecoprice reduces salaries, perhaps below the level where workers can reproduce their labor power. If the production of a given commodity continues, thanks to nonecological prices and salaries, this means that not only do the capitalists exploit their workers, by claiming all the surplus value, and impoverish future generations through extraction of limited resources, but that even these workers impoverish future generations.

This hypothesis can be made more concrete by examples, among them cutting down and selling palm trees in an oasis. There is a limited number of trees, and one may either limit the cutting to what nature reproduces, possibly with the support of cultivation and replanting by man, or cut the trees down more rapidly than they grow back. In either case, the activity may be profitable, in terms of the market economy. But the activity is only ecologically healthy in those cases where the felling does not exceed the reproduction. And by calculating what will have to be added to the price to cover the labor necessary for the different degrees of replanting, we shall see different profitabilities, in the ecological sense. In the cases where there is neither profit nor salaries, if the palm trees that have been cut down and sold have been replanted, and if the expenses of replanting have been added to the prices, both the capitalists and the proletarians live parasitically from the extraction of a limited resource (an extraction that is harmful to future generations).

This example is obviously abstract, a single activity having been isolated from the rest of a market economy. Here is another, more realistic example: The extraction of petroleum in Kuwait procures huge profits for the state and a considerable profit for the oil companies. At the same time, the salaries of the workers in the petroleum sector are several times higher than average salaries in the Near East. And labor in the oil firms is no harder, physically, than other kinds of labor in the region. Is it right to say that these worker create all these values, both the enormous profits and the high salaries? To the response that this is a situation where the prices are raised by monopolies, one may object that oil products are, on the contrary, too cheap, given what they would cost if the ecological damages where added to their prices. Let me put it this way: The Marxian theory of value founded upon labor is valid for the *reproductive* forms of production. But in an *extractive* form of production, value is transferred from resources to profits, which may then be called an extractive surplus profit. This extractive surplus profit can be so large that the entire production process, at all levels, can receive more value from it than the labor itself has created.

In this case, one must not say that every profit comes from the underpayment of the labor included in the product. According to the formula "profit equals sales price minus cost price," one can increase profit by raising the sales value, the price, or by lowering the cost price, through lower salaries, longer workdays, increased productivity, automatization, rationalization, through a monopoly on raw materials, through a decrease in social expenditures of various sorts, and through omission of expenditures for ecological restoration.

But who is "exploited" by this extractive surplus profit? Nature, and, indirectly, future generations. Extractive surplus profit represents future pauperization. Oil companies, for example, are not only exempted from accounting for current ecological expenditures, they also omit future expenditures by depriving future generations of vital resources. (The idea of a future "exploitation" presupposes, among other things, that resources are limited and that the population is growing, not to mention that petroleum is a resource that can be transformed into a food for human beings that is rich in protein, something that will necessarily be lacking in the future.)

Here we can anticipate all sorts of objections. For example, the ecological price, stipulated on the market, is problematic, quite simply because it is a market price. One cannot blame workers for the inadequacies of capitalism, since even the workers' consumption is a part of the capitalist economy. Despite all the false needs introduced by the system, the desire to obtain a higher standard of living through higher salaries is a good and fair thing, even when the supposedly higher standard of living, measured by gross national product, mainly represents a commercialization of activities that until now have not been recorded by the market, or even if today what we record as an apparent improvement of the standard of living represents activities introduced to compensate for environmental destruction. It is, therefore, not appropriate to conclude by preaching to so-called consumers, that is, the workers, that salaries and consumption should be reduced. Production itself implies consumption, and consumption implies production by the worker; if one wishes to change production and consumption, this cannot be done by isolated consumers who have been induced to cut down by moralistic pressure, but only by better organized producers: capitalists, or trade unionists, or labor parties.

However, the dilemma remains. When the economy is viewed in an ecological perspective, the need for an anticapitalist politics appears obvious, and such a politics must be linked to trade unionists and worker parties; but, at the same time, the organized struggle of classes for better salaries has become ecologically problematic. The struggle for better salaries is today insufficient, just as a planned economy is necessary but insufficient. What is needed is greater solidarity, geographically and historically; ecological knowledge must penetrate the masses, managers, technicians, trade unions, and political parties. This problem is essential for Marxist intellectuals and political militants, today.

In the same perspective, the time has come to reconsider some central conceptions of Marxism, particularly the *theory of cyclical crisis, the theory of pauperization,* and the *theory of revolution.* I will discuss these briefly. We notice a growing pauperization, especially in the less-developed countries, and various forms of relative impoverishment among the peoples of the rich countries. From this one can foresee the possibility of a certain polarization and an intensified class struggle. But the ecocrisis probably affects different groups in different ways, and at different times. Consequently, it will probably be very difficult to establish a common basis for global proletarian solidarity. Thus, the danger of a certain fascistic tendency among well-paid workers in rich countries is a real danger. The concepts of crisis, impoverishment, and revolution are important concepts, which point to the imminent self-destruction of the capitalist system, but the content of these concepts needs to be rethought.

The concept of the working class also needs to be reexamined. Even if an unskilled worker in Detroit and a farmworker in India are both deprived of the means of production, and even if both of them produce surplus value for others, nonetheless, there are material reasons why the American worker and the Indian worker consider each other as adversaries rather than as comrades. To put it simply, one eats the bread that the other does not have the means to buy for himself. If the concept of "worker" is essentially defined by the relationships of production, the concept is inadequate in most of the political contexts where different groups of workers have very diverse possibilities for satisfying their needs. This definition is only adequate if proletarianization at the same time implies that all proletarians have more or less the same material situation—in any case, after a certain period of time. But in the ecological perspective, it is doubtful that this hypothesis will be valid: the poor in the Third World will never reach the level of consumption of a skilled worker in Europe or the United States.

Thus, if the existing conditions of nature are introduced into the infrastructure, the concept of the working class must also be defined in relation not only to the means of production, but also to the conditions of nature.

The moral conscience of the eighteenth-century bourgeoisie could tolerate the material inequality of the period because of their belief in progress: in the future, everyone will have a comfortable life. Up to this point, the socialist movement has regarded the current inequality between rich and poor workers in just this way: there has been a belief in progress and growth; in the future, in any case, when (and if) socialism triumphs, we shall all live well. But what if the future, socialist or not, implies the continuation of a growing differentiation between workers who live well and poorly nourished workers?

How can the workers in an underdeveloped country influence international capitalism? For example, can the they use the strike as a weapon against the centers of capitalist decision making located abroad? Perhaps the oil policy of the Arabs last winter will serve as an example to underdeveloped countries that export raw materials: by uniting against the developed countries, they can win a greater control over their own resources and secure better prices for their products. For certain underdeveloped countries, this policy may prove fruitful. But it remains to be seen whether this sort of policy will improve the lot of the masses in these countries, or only that of the elites. In this regard, the Arab example is not entirely convincing.

At an abstract level, everyone certainly has a common interest, that of avoiding an ecocatastrophe. This point of view is also important for the concept of class struggle. Before the threat of global destruction, all of us—capitalists or proletarians, rich or poor—have a common interest in survival. But the concrete meaning of this common interest is not the same for all.

If one does not seek to resolve the ecocrisis in an egalitarian way, but in an ethnocentric fashion, the struggle (even if it is not a class struggle in the orthodox sense) will probably intensify.

As for existing socialism itself, two essential changes will in any case be necessary for the classic model of the Soviet Union: a passage from economic growth, in the traditional sense, to ecological stability from extraction to reproduction, preferably by autonomous and self-supplied communities; and the passage from a pan-technological enthusiasm to a greater appreciation of agriculture relative to industry.

This change toward a productive and agricultural socialism implies changes in our conception of what is progressive and what is not. It also implies changing certain fundamental values and attitudes: a respectful, rather than an aggressive and exploitative, attitude toward the laws of nature; and an emphasis upon cultural, social, and political values, instead of an exaltation of economic growth in the traditional sense.

Virtues such as solidarity and discipline will be called for: neither selfishness nor ethnocentrism, but an egalitarian way of life; neither waste nor exploitation, but labor and prudence. And perhaps, in this perspective an "oiko-logical" thinker, like Plato or Aristotle, may have something to teach us about the ideal life, both before and after the revolution.

These are a few essential themes of Marxist theory and practice that need to be rethought. Of course, it is not simply a matter of introducing a few more or less new concepts into the "Marxist system." It is a matter of concretely rethinking theory and practice, in regard to ecological problems, and this reexamination must be conducted with the participation of all the members of the Marxist movement, theoreticians and politically committed individuals, in different situations and places.

NOTE

1. Marx *Das Kapital,* in *Okonomische Schriften,* ed. Erster Band (Stuttgart, Cotta Verlag, 1971), 594–596. Translation taken from Marx, *Capital,* trans. Ben Fowkes (New York: Penguin, 1976), 1:637–638.

Chapter 7

MARXISM AND THE ENVIRONMENTAL QUESTION

From the Critical Theory of Production to an Environmental Rationality for Sustainable Development

ENRIQUE LEFF

INTRODUCTION

It might appear "antiparadigmatic" today to pose the environmental question from a Marxist point of view and to reformulate Marxism from an environmentalist perspective. The collapse of really existing socialism and the international consensus in favor of a strengthened new global order based on a market economy have also undermined the legitimacy of the Marxist theory of history and political economy—a theory that, despite having generated critical analysis concerning the causes of environmental problems that arise from capital accumulation and economic rationality, has not integrated nature (or ecological processes) into the general conditions and process of production.

Marxism in fact offers the theoretical basis needed to demystify the dominant neoliberal discourse and to clarify the current conflict between the conditions of sustainable capitalism (based on the expansion of investment, production, markets, and profits) and those of ecological and environmental sustainability. Yet the discourse of modernity has led to the delegitimization and abandonment of Marxism as a critical theory of society and the environment. In the acritical discourse of sustainable development and "natural capital," environmental issues are considered part of the "new economic order" and of the global transition to liberal democracy. Although these issues have generated a social response, visible in the new environmental movements, the latter still lack a theoretical framework and a strategic program for the construction of an ecologically sustainable productive rationality.

The ecological vacuum of conventional economic theories and historical materialism is increasingly evident and has given rise to several strains of analysis and reflection regarding the effects of economic growth on environmental degradation and the destruction of natural resources: the difficulty of quantifying natural and cultural wealth, as well as long-term economic, social, and ecological processes, in terms of market values or prices; the irrationality of capital accumulation from the point of view of energy degradation and natural resource depletion; and the inability to control and internalize negative impacts on nature and society by using economic concepts and instruments.

Starting with these questions, we can begin to reconstruct a Marxist theory of production that accounts for the incorporation of natural processes in the general conditions of production and the construction of an environmental rationality based upon the principles of ecotechnological productivity, participatory environmental management, and ecological sustainable development. Whereas Marxism offers a historical, economic, and social perspective on the study of environmental problems, and a theoretical paradigm that can be reworked in a way to incorporate the environment into the productive process, the environmental perspective offers Marxism knowledge about the socioenvironmental and ecological conditions of sustainable development.

This chapter will treat some of the challenges that environmentalism poses to Marxist political economy and the paths opened up for an ecosocial theory of production. I stress that historical materialism, as a critical theory of history and political economy, must rework its categories of nature and culture, placing them in the very center of the productive process. It also must question the capitalist and socialist models of economic growth based on maximization of profits and short-term economic surplus. I will suggest a new theory of production (a new productive rationality) that incorporates the environment into the productive process not as externalities of the productive system nor only as a general condition for sustainable production, but rather *as potential for an alternative productive rationality.* This provides a new orientation toward, and basis in, the development of the forces of production that integrate cultural, technological, and ecological processes, thus generating productive processes that are equitable and sustainable. These questions will be posed from a political and a theoretical perspective, that is, considering the environmental potential of sustainable development and the political power of social movements in the construction of environmental rationality.

HISTORICAL MATERIALISM
AND THE ENVIRONMENTAL CRISIS

How does Marxism deal with the environmental problems of our time? Although Marx could not predict the magnitude of the current environmental crisis and the extent of the global ecological imbalance, he did anticipate the effects of the capitalist mode of production on the destruction of planetary resources and the loss of soil fertility. But Marx's account of the effect of production on the destruction of nature did not generate a theoretical response, or an internal criticism of Marxist concepts. Nature was the object of reflection by Marxism insofar as it was seen as a superstructural formation produced by various cultures at different historical moments. Marxism, in order to establish itself as a social science, had to separate itself from the naturalist ideology of its time, leading to the fertile development of the theory of social determination over nature, which included nature's incorporation in the sciences and technology but excluded its specific contribution to the creation of wealth. It was only the ecological crisis that has become manifest over the past twenty-five years that set off an incipient reflective process within Marxism concerning its theoretical concepts and epistemological bases, giving rise to a critical consciousness that today coincides with the collapse of really existing socialism.

Environmental discourse during the 1970s was regarded skeptically by some self-proclaimed Marxist theorists as "false consciousness"[1] regarding the fundamental causes of the economic crisis, behind which lay the need to open up new productive sectors (e.g., the antipollution industry) to provide an outlet to reinvest profits.[2] The environmental issue itself, the need to preserve natural resources and ecological balances, the basis for all sustainable economic and productive processes, was almost completely ignored.

The unmasking of environmentalists' "false consciousness" still conceals one of the basic notions of historical materialism's analysis of capitalism and of the building of socialism: that the transparency of social relations and of the society–nature relationship would come about only as a result of the elimination of the scarcity that characterizes the precapitalist and capitalist modes of production. Therefore, although the transition to socialism would be led by class struggle, its objective possibilities would depend upon the development of the forces of production and, in particular, on the forces of nature released by the scientific–technological revolution—thus the will to permit the unrestrained development of productive forces, both under capitalism and during the transition and building of socialism (in the latter case, under different forms of ownership and control of the means of production), both of which have provoked various types of environmental deterioration. This ideological inertia has ignored the ecological limits of growth and the ecological bases for a lasting and sustainable development of the forces of production, which was recognized by Marxist theoreticians only in the late 1980s.[3]

The crisis of Marxism is not only the result of the collapse of really existing socialism and its self-destructive development of the forces of production. It can also be attributed to certain blind spots in historical materialism as a science of history. Although Marxism understood capital as a relationship of exploitation and promised the rationalization of the development of the forces of production by eliminating internal crises of overproduction and profit making, it did not incorporate the forces of nature, defined as productive potential, in its theory of social labor.

Marx turned Hegelian dialectics and grounded contradiction in the social relations of production. But he could not escape from the state of knowledge of his time, and he searched for a unit of value quantification based on the simple and direct labor to which capital reduces labor power. The concept of socially necessary labor time, upon which value is measured, is devised using capital's exploitative relationship with labor. However, this measurement of value excludes the forces and conditions of the production of nature, which in their diversity and ecosystemic complexity are irreducible to homogeneous units. Marx's theoretical basis thus is put into question not only as a result of ecological destruction and capital's inability to assign a value to natural resources (critical problems that remained external to the purpose of *Capital*), but also because of the indeterminate effects introduced by technological progress in the calculation of socially necessary labor time as a measure of value.[4]

Marx's theory of production does not incorporate natural and cultural conditions that participate in the production of value. Also, it is incapable of putting a value on natural and cultural resources. The environment question thus forces Marxism to reformulate its teleological vision of history, which is based upon a one-dimensional development of productive forces. The environmental question also poses the need to respond to present-day world transformations and power relations in which new social relations and historical subjects emerge, creating new conditions for social labor.

The need to incorporate ecological processes into the analysis of production is opening up a series of theoretical developments and methodological approaches. One problem that arises is whether or not the processes that determine the formation of natural resources and the ecological conditions of productive processes are the object of another science, ecology, which is external to historical materialism, understood as a critical theory of the capitalist mode of production. This problem opens up two options. The first is the possible articulation of two sciences, ecology and historical materialism, in which ecology would account for the structure and workings of the ecosystemic base of natural resources, or of the constraints, norms, and ecological support for the production process, which are external to capital's internal contradictions.[5] The second is to see the environment as the articulation of different productive processes (natural, cultural, economic, and technological) and ecological processes as codeterminant processes of production, leading to a paradigmatic reformulation not only of the theory of capitalist production but also of all sustainable development processes.[6]

This reconstruction of the theory of production, integrating social and natural processes, goes beyond a Marxist concept of nature understood as a mediator in all processes of the social appropriation of nature.[7] In fact, some authors have tried to rescue an "ecological" Marx based on the philosophical discourse on the relationship between society and nature, and to uncover a second, nonexplicit contradiction in his theoretical discourse. Nevertheless, in Marx, nature appears less as a second contradiction of capital than as a "secondary contradiction," an effect that is overdetermined by the exploitation of labor.

General categories of nature and labor, however, do not permit one to capture the specificity of the relations between a particular social formation and its environment. Yet the concept of a socioeconomic formation opened up an important way of studying precapitalist societies and even allowed one to think about the relationship between cultural organization (e.g., family, marriage, and kinship relations; religious representations and ideological formations) and the natural environment in the development of productive process and resource use patterns.[8] These contributions by Marxist theorists are important, but they provide an insufficient theoretical basis for analyzing the ecological and cultural processes affecting today's sustainable development of productive forces. From an environmentalist perspective, we should not see culture only as superstructural values or as a structure that takes the place of the productive base in traditional, noncapitalist societies; rather, we should translate cultural values and organization into a principle of productivity in the sustainable use of natural resources.[9]

Recently we have seen the emergence of an eco-Marxist current that emphasizes nature's function in supporting or limiting production. But because it lacks a theory explaining the transition toward, and construction of, socialism based on environmental rationality, this approach does not incorporate natural processes into the productive process itself, leaving intact the paradigm of the capitalist mode of production. Therefore, there is a need to establish a concept of nature that is appropriate for the building of socialism based on the social use and democratic and participatory management of the environment viewed as a resource base, means of production, and condition of existence, which in turn determines different production life-style patterns.[10]

The environment is more than just an element of the conditions of production or a cost of economic growth. We must see it as productive potential, as part of the social forces of production, in a productive paradigm that is not economistic yet pertains to political economy, since the environmental rationality of production involves environmental management that includes social participation regarding resource appropriation. The construction of environmental rationality will permit a resolution of the contradictions between conservation and growth, between environment and development, between the self-destructive appropriation of nature by capital and the aim of subsuming the appraisal of the environment under the concept of natural capital.

Environmental management combines the political conditions of direct democracy with conditions of equitable and sustainable production. Environmental democracy is not only counterposed against state bureaucratization but also offers a way in which the environmental movement participates in the deconstruction of market mechanisms and the creation of a new productive rationality.[11]

In this sense, the environmental challenge to Marxist theory goes beyond a synthesis of orthodox Marxism and the new ecology, or the incorporation of energy rationality into the metabolism of production. The aim of eco-Marxism is to integrate new principles into the development of the forces of production and the democratization of society through a theory that encompasses not only economic costs and externalities but nature as a force of production, as productive potential. That requires the elaboration of new concepts concerning an equitable and sustainable production process along with instruments for planning, controlling, and monitoring the environment (through environmental impact assessment, integration of new indicators of sustainability into economic accounting, methodologies for territorial layout, ecological bases for sustainable agriculture, etc.). At the same time, the political process is opened up for greater participation in decision making by civil society.

This environmentalist perspective offers socialism with a human face and an ecological base, and a democratic transition toward a new productive rationality based on the principles of participatory management of productive resources via the socialization of the means of production, understood to include natural processes and cultural resources.

MARXISM AND ENVIRONMENTAL CONDITIONS OF PRODUCTION: FROM CRITICAL ANALYSIS TO PURPOSEFUL DISCOURSE

Increased production necessarily brings about an increase in the productive consumption of raw materials, a heightened rhythm of extraction and transformation of inputs, and the production of waste. But the failure to put a value on natural resources means that ecological imbalances, the decline in soil fertility, and the depletion of nonrenewable resources are not reflected in the value of capital and in price formation of natural use values, while the accumulation and reappraisal of capital is reflected in the destruction of resources that capital does not take into account.

The environmental crisis thus marks the explanatory limits of a theory in which natural use values are valued only insofar as they incorporate labor time or internalize the "scarcity" of natural resources through the market. Social labor must be redefined to mean necessary labor for production and renewal of production conditions in the framework of a new and changing international division of labor, the scientific-technological revolution, the enhancement or degradation of environmental potential, and global changes. The question of the environment challenges the theoretical status of the concept of social labor and the concept of value not only in the widened reproduction of capital or in the restricted fields of environmental protection practices and the capitalization of nature, but for the construction of a productive paradigm based on environmental rationality within the context of which production and the productivity of social labor are bound together with the ecological conditions for sustainable and lasting production (not for the production and reproduction of capital).

Marxism is not merely enriched with environmental concepts. At the same time, it contributes to an environmentally critical and positive theory of production. First, Marxism posits the social and political character of sustainable development and offers a theoretical framework within which to analyze the connections among various processes, the links between different conditions of production—the "objective" and "subjective" conditions for the construction of a paradigm of sustainable development—and an analysis of the new environmental movements not only in relation to their contribution to a postmodern culture of difference, but also with respect to the opposition and conjuncture of political and class interests in the construction of a sustainable and lasting productive process.

In this conceptual exchange between environmentalism and Marxism, we see new possibilities for a critical analysis of the relationship between society and nature—and between the economy and resources—and for an eco-Marxist theory of production.

NATURAL AND ECOLOGICAL LIMITS OF PRODUCTION

An analysis of nature as a limit or norm for production brings us to the question of limits on growth under current conditions of capital expansion and technological change.[12] It also leads us to analyze, on the basis of natural resource requirements and the possibilities for recycling and diluting production and consumption waste, the impact of this expansion on resources and the global ecological equilibrium. Although the environmental crisis is manifested generally in local impacts and not as an absolute limit on growth,[13] this crisis is becoming increasingly global (e.g., ecological imbalance, food and nutrition crisis, poverty, etc.) with transnational and cross-class effects. This situation invites us to question the concept of scarcity within the framework of new strategies of capital (e.g., technical innovation, new products, market expansion, and environmental protection).

At the same time, these limits should be reconsidered in the light of different strategies of resource use. Proposals for zero economic growth were widely rejected because these were not viable alternatives to halt capital accumulation, and also because Third World countries have the right to develop. It was in this context that environmental principles were put forth as an alternative, nondestructive road to development, which would be responsible for social equity today and would preserve natural resources for future generations.[14] Another productive paradigm was considered, one that alters the limits of the development of productive forces by transforming the premises, principles, and development of the forces of production.[15]

NATURAL PROCESSES AND THE GENERAL CONDITIONS OF PRODUCTION

The general conditions of production, understood as everything that exists in nature and society that is not produced according to the process of value formation and the laws of the market but rather that establishes the necessary conditions for capitalist production, should also be redefined in the current context of widened capital reproduction. In this sense, it is not only ecological conservation that needs to be added to the economic policies of the state—policies that develop productive areas, provide services, and supervise all those activities considered to be of strategic value for the state and that are unviable for capital (e.g., infrastructure and public services, basic foodstuffs, and health services). Traditional public sectors are being challenged by neoliberal economics and by really existing capitalism through privatization policies that are transforming the conditions of production throughout the North and the South, East, and West.

What needs to be elaborated are those conditions for production that are most difficult for capital to generate and regenerate, those processes that are excluded from economic rationality due to their inability to be quantified in the short run and also because they cannot be valued in capital terms: the natural base and the ecological supply of resources; conservation conditions and the regeneration of natural resources; environmental services and common goods; health conditions, environmental quality, and the quality of life; long-term ecological processes and their global and transgenerational effects; and people's natural and cultural patrimony. The environmental conditions of production would include all these processes that cannot be absorbed under the concept of natural capital.

We thus need to work out a theory of those social and natural processes that intervene to provide the *ecological conditions of production*: on the one hand, the state, by establishing a system of protected areas and a normative juridical system regarding the ecological order of productive processes, spacial distribution of productive processes, industrial and domestic waste management, and so on; on the other hand, civil society and communities, through a series of productive and consumer practices that are nonpolluting, many of them located outside the realm of capital and the market. Self-managing environmental units of production would supply basic needs to communities and (as a result of environmental conservation) contribute to the maintenance of the general conditions of the production of capital, the productive conservation of resources, present and future social equality, and sustainable and lasting development.

ENVIRONMENTAL EXTERNALITIES AND ECOLOGICAL COSTS OF PRODUCTION

Ecological destruction, the overexploitation of natural resources, and environmental decay have been the effect of the capital accumulation process. Environmental resources are a condition of production that also appear at present as a cost for the widened reproduction of capital. Nevertheless, these ecological and social costs continue to be externalized to the environment. Two moments of capital have combined in this process. On the one hand, the failure to place a value on nature favors the overuse of resources; also, the widened accumulation of capital generates expanded demand for natural resources that enter into the productive consumption of capital. On the other hand, realization crises and technological innovations generate processes of destructive production, multiplying and accelerating the extraction of nonrenewable resources and imposing productive patterns that are aimed at maximizing short-term profits without taking into consideration the conservation and regeneration of natural resources.[16] This process clearly leads to further deterioration of the environment and the quality of life.

Capital's externalities thus become new production costs. These are mobilized by political pressure and are not established by any economic mechanism of price formation. In response, the internalization of socioenvironmental externalities and the introduction of necessary ecological costs has been proposed in order to maintain capital earnings. Thus, the neoclassical approach proposes the concept of "natural capital," which includes nature and the environment within the realm of capital. The task is not an easy one given the codependence, incommensurability, and externality of the environment and the economy.[17] There are limits on capital's (and the state's) ability to translate nature into market (or planning) prices, especially of ecological processes such as the resilience, regeneration, and recovery of ecosystems in the face of capital intervention, as well as nature's capacity and potential to contribute to the production of use values. Labor value, as a concept that binds together the labor process and the capital cycles, cannot be easily tripolarized in order to internalize the natural processes that contribute to the production of use values and exchange values. Nature cannot be disaggregated into discrete and homogeneous value units, like labor versus capital, nor can it assimilate the multiple natural cycles to capital cycles.

Therefore, it is not enough to regard nature as a cost that is calculated in terms of natural capital. We need a political economy of the environment that is critical, in order to see poverty, unemployment, and the destruction of natural resources as effects of given relations of production. But we also need a *positive theory of production*, which can give support to a new productive rationality in which natural forces along with labor power assist in the development of productive forces and contribute to the production and distribution of wealth, social equality, ethnic diversity, ecological balance—and to a sustainable supply of natural resources.

DIFFERENTIATED ENVIRONMENTAL EFFECTS AND ECOLOGICAL COSTS OF UNEVEN AND COMBINED DEVELOPMENT

There has been a rich discussion about the impact of capital on different environments and socioeconomic formations. In contrast to the ideological discourse on global environmental problems and humanity's common responsibility, theoretical and case studies within the field of uneven development emphasize the difference between the environmental problems of rich and poor countries, as well as the uneven distribution of the environmental costs among nations, regions, and social classes generated by the uneven exchange between the North and the South, leading to overextraction of resources and the transfer of destructive and polluting processes to tropical countries.

This allows us to see the combined effects of environmental degradation and social polarization, as in the case of the decline in soil fertility and resources owing to deforestation, erosion, salinization, and desertification caused by the introduction of inappropriate technical models (e.g., capital-intensive agriculture, monoculture, and cattle raising in tropical regions) and their impact on poverty and malnutrition. Another example would be urban-industrial concentration, rural–urban separation, and regional migration, all of which can cause serious environmental problems.[18] The occupation of space and the appraisal of resources that result from uneven development have brought about the overexploitation and underutilization of environmental resources, and a failure to take advantage of the environmental potential for ecologically sustainable development.

The negative effects of uneven and combined development should be counteracted by an alternative mode of production capable of combining multiple strategies of sustainable development, and even of exploring possible complementarities between market rationality and the growth of self-sufficient environmental economies.[19]

THE EMERGENCE OF POLITICAL ECOLOGY AND ECOLOGICAL MARXISM

Political ecology and ecological Marxism are new areas of action and thought that began outside historical materialism, although they borrowed Marxism's principles of social criticism to apply to a nonecological analysis of current environmental problems. Without establishing a clearly constituted theoretical paradigm or scientific discipline, they seek the causes, roots, and social effects of the processes of ecological imbalance and decay. At the same time, they suggest principles for the ecological and energetic reordering of production. In this way, a new energy economics has been constituted that provides a critical theory of the energy degradation produced by economic rationality.[20] In addition, new anthropological approaches analyze the rationality of traditional societies in terms of their energy flows.[21] At the same time, the emerging field of ecological economics establishes both global conditions for all economies and the ecological bases for subsistence microeconomies. However, these contributions cannot replace political economy as a specific field of social production, in which one must incorporate the ecological and energy bases, conditions, and potentials for an equitable, sustainable, and lasting production process.

Within this tendency, an eco-Marxism inspired by the thermodynamic theory of open systems has also appeared. In opposition to the unified, uniform, and quantitative modern rationality, based on the predictability, normative character, and control of natural, cultural, and social processes, eco-Marxism posits concepts of opening, diversity, indeterminacy, coevolution, codependence, and dispersion.[22]

There are thus several ways of enriching Marxism in order to develop a critical analysis of production through the consideration of environmental issues. Along with those noted above, I mention three more, which I believe to be fundamental:

First, the constitution of socioenvironmental formations as units of production: Marxism offers a holistic vision of the relationship between society and nature based on social rather than biological overdetermination and the centrality of production and social reproduction. From an environmental perspective, the objective is to incorporate ecological bases and conditions into the capitalist production process and then to analyze the articulation with environmental units of production, that is, with all socioeconomic formations that are not strictly capitalist. In its most general form, these virtual economic–environmental formations can be included within a broad typology of social formations depending upon different forms of land ownership, social access to resources, property in the means of production, and the market exchange of products and inputs.[23] Nevertheless, these typologies do not take into account different links with nature, for example, appraisal of natural resources, environmental services, and ecological potentials in the productive process. Environmental management thus poses the problem of articulating various self-sufficient economies and environmental units with the market and with different ethnic styles of resource use.

Second, a social analysis of the global nature of, and interconnection among, socioenvironmental processes: Marxism is a social theory, based upon the material nature of the social relations of production, which establishes the relations of determination, causality, and conditioning with a series of natural and social processes. It is a theory (the first one, still valid today) that denaturalizes and desubjectivizes social processes.[24] In this sense, Marxism opposes naturalist, biological, and energy-centered approaches, as well as methodological individualism, all of which analyze the relationship between society and nature from the point of view of biological evolution or from the ecological perspective of the carrying capacity of growing populations in different ecosystems. Placing the society–nature relationship within the social relations of production politicizes the question of the environment, which appears as a complex and polyvalent object that permits the transformation of the dominant economic rationality and the construction of a productive rationality based upon the articulation of natural, technological and social processes.[25]

Eco-Marxism thus allows us to think about complexity, but not in the manner of physical constructivism, generalized ecology, or general systems theory.[26] Rather, it moves from an analysis of the complex whole based on the capital–labor relationship to the study of complex socioenvironmental systems and economic paradigms based on production relations and labor processes that integrate cultural values, ecological productivity, and technological progress as processes that are linked to the development of the forces of production. One must go beyond systemic approaches that try to incorporate the environment to the economic system, through planning practices,[27] in order to study coevolutionary processes between the economy and geoenvironmental systems.[28]

The environment should not be subsumed under the capitalization of nature nor maintained as a system external to the economic sphere, but rather it should be integrated into production not only as a production condition, but also as a productive force. The environment should be regarded as the articulation of cultural, ecological, technological, and economic processes that come together to generate a complex, balanced, and sustained productive system open to a variety of options and development styles.

Marxist theory is able to analyze the relations between productive overdetermination and a variety of processes (e.g., political, juridical, natural) based on the fundamental opposition between capital and labor. From an environmental perspective, the social relations of production are made more complex, lying between capital, labor, and natural processes. The durability of the productive process is no longer established through programs of economic recovery and market expansion. Rather, it stems from a dialectic in which the ecological conditions for the renovation and productivity of natural resources, global ecological balance, territorial distribution of productive activities, and the participatory management of resources are fundamental for attaining a sustained and equitable development.

In this way, the environment is seen as a complex system that articulates natural, technological, and cultural processes within the social forces of production, and which is interlinked with new social relations of production—relations between civil society, state, and nature, and also relations between communities and their environment as means of labor and production for the participatory management of their natural resources.

Third, a change of productive paradigm that includes nature and culture as productive forces: Returning to the questions that I raised earlier, I propose an alternative productive paradigm to that of the dominant economic rationality, one in which production would not primarily depend on the development of productive forces and technological progress propelled by the logic of the market and profit maximization but on an increase in the production of use value to satisfy socially and culturally defined necessities, based on the socialization of access to nature, the decentralization and ecological planning of productive activities, and the management of people's and communities' environmental resources. This reconstruction of productive rationality stems from the ecologically normative nature of the global economy and especially from the development of regional microeconomies.[29] In this paradigm, nature and culture are not just mediating processes but act as social labor and direct productive forces. In this sense, I am proposing an ecotechnological paradigm for the integrated, sustainable, and lasting management of natural resources, based upon three articulated levels of productivity.[30] The first is the level of ecological productivity, derived from nature's potential (ecosystemic organization, photosynthetic process, etc.), which generates a system of natural resources with a growing and sustainable supply of natural use values. The second is the level of cultural productivity, in which the diversity of cultural organizations and ethnic identities is converted into a productive and social force, and which recuperates and improves their traditional practices to become sustainably productive, as well as incorporating technologies so that communities can manage their own environmental resources. The third is the level of technological productivity that is based on a complex and polyvalent technological system capable of driving the previous levels without destroying the bases for ecological sustainability and cultural diversity.

ECONOMIC RATIONALITY AND ENVIRONMENTAL RATIONALITY FROM THE PERSPECTIVE OF SUSTAINABLE DEVELOPMENT

The environmental issue has presented economics with new problems that are neither perceptible nor resolvable through traditional paradigms.

From the perspective of sustainable and lasting development, one can clearly see the ecological and energetic irrationality of economic growth, fueled by the maximization of private profits and short-term economic surplus, that, in turn, generate the economic system's negative environmental externalities. Environmentalism poses ethical, political, and productive solutions for the construction of new development styles based on economic decentralization, ecological productivity, and political pluralism.

As opposed to the normalization of productive patterns and the homogeneity of life-styles under the unifying logic of the market, environmentalism proposes an alternative based on the ecological diversity of nature and the cultural diversity of peoples. This means it is impossible to compare the one-dimensional neoliberal model with a totalizing socioenvironmentally defined rationality. The multiplicity of development styles possible within an environmental rationality is the result of diverse ecological conditions of time and space, but more importantly, of the actions of "ecological interests" that can mobilize a wide range of political actors and social groups in the reorganization of productive rationality.[31]

Capitalist economy lacks instruments with which to appraise the contribution of ecological and natural processes with respect to the production of natural use values and products. Economic calculation is unable to place a value on long-term ecological and social processes.[32] Despite efforts to create a concept of natural capital that can internalize environmental externalities within economic calculations, the market economy is unable to demonstrate rational criteria for the investment of limited resources. The problem is intensified by the uncertainty of technological changes (insofar as their effect on opportunity costs are concerned), the viability of using different potential resources, the rhythms of substitution of the resources (the incorporation of natural wealth into production and market circuits), and impacts on local and global environments.[33]

Many of the values and qualitative goals of environmentalism (e.g., conservation, ecological potential, political pluralism, ethnic diversity, aesthetic values, direct and participatory democracy, and quality of life) are incommensurable; they cannot be reduced to a common measure. William Kapp warned that the comparative evaluation of economic, energy, and environmental rationality requires essentially heterogeneous units of measurement, for which there is no common denominator.[34] Environmentally sustainable development requires new analytical concepts and instruments to evaluate the patrimony of natural resources, ecotechnological productivity, and self-managed subsistence economies aimed at the sustainable and lasting production of use values and market commodities.

We are then presented with the articulation of new spaces of environmental management (not oriented directly toward the production of value nor subject to market laws) with an expanding capitalist economy. Two different rationalities confront one another. Not only do the preferences of future consumers come into play; so do the various "ecological interests" and rationalities of current consumers and producers, with alternative strategies for property, possession, appropriation, transformation, and usufruct of environmental resources, situated in an economic and political field that is open to complementarity and negotiation but not free from contradictions and sociopolitical struggles.[35] This is a field that is open to eco-Marxist theory.

POLITICAL CONDITIONS
FOR SUSTAINABLE DEVELOPMENT:
THE ENVIRONMENTAL MOVEMENT

The crises of socialism and the environment have opened up new ways of thinking about transforming the world. The new social movements are oriented toward democracy based on an environmental culture; they reject the idea of a historic transformation led only by the working class, aiming at the collapse of capitalism (peacefully or by revolutionary means) and the construction of socialism. The aim is to construct new styles of development and new models of civilization that permit sustainable, lasting, and equitable development, and challenge not only economic rationality but also bureaucracies, in ways that encourage political pluralism and the participation by civil society in the management of its productive and vital processes.

The aim of such a transformation is no longer just the ownership of the means of production, which ensures a more accelerated development of the forces of production and a better distribution of wealth, but, additionally, the conservation and enhancement of the resource base for sustainable development, that is, the conditions of existence and labor, distribution of earnings, and the improvement of the quality of life and the environment through a radical critique of needs. It is a question of conserving common goods, of recovering the environment as productive potential, and as a means of production and a means for life. New struggles are thus being waged around ways of appropriating and socializing nature.

Environmental thought can be inscribed within post-Marxist or post-modernist thought. It defends the specificity of local action (thinking globally, acting locally), the autonomy of social groups, and difference—difference in cultural values, and development styles, and options. It presents new ethical values and a new political culture, but at the same time poses the problem of political efficacy and of the real political power held by environmental groups. Although the defense of autonomy and local difference can be seen as part of the struggle against totalitarianism (against vertical and corporative power structures in traditional political organizations), it also rules out any universal demand other than one claiming the legitimacy of all local demands. Nevertheless, the demand for autonomous spaces as a starting point for the development of alternative local productive projects is part of the larger movement for another kind of material existence based on the integration of multiple development styles. This demand for plurality goes beyond the democratization of the state through political representation, which in many cases is creating new forms of authoritarian regimes imposing a unidimensional neoliberal rationality on the people.[36]

The environmental movement, driven by the principles of participatory management, the rights of ethnic groups and peasant communities over their territory, and the reappraisal of aesthetic values in nature, is generating new forms of participation in the social process of production and the socialization of the environment as production conditions.

In contrast to orthodox Marxism's concept of a class—the proletariat—as the protagonist of social change, the environmental movement presents itself as an explosion of interests and identities that rejects a unitary class response to capitalism as a mode of production. Nevertheless, this movement must still redefine a socialist project that can integrate its particular differences into a strategy based on common social labor, one that can respond to "post-Marxist" positions, which themselves have been unable to find unity in fragmented social identities.

Unified objectives are clearly fundamental in order to build alliances that allow for a strong environmental movement with real power. The question is whether the unity/force of the movement should be based on the unity of labor in its dialectical relationship with capital; or whether the dynamic is more complex, meaning that the objective of environmentalism would be the conjunction of different interests and identities within common aims—not the collapse of capitalism for the construction of really existing socialism, but, instead, the construction of a new *environmental social rationality* that allows diverse styles of life and sustainable development.

In this sense, "the challenge of specificity is accepted by all the new social actors. . . . It is the result of the complex network of policies . . . implemented by capital and the state in order to integrate people at the same time as production conditions are changed. On the one hand, this specificity (difference) represents a break with collective and class solidarity. On the other hand, it reveals new micro-networks of social solidarity and a universal network of solidarity based on social citizenship."[37]

Clearly, "questions relative to the conditions of production are questions of class, although they may be more than questions of class."[38] This means that, despite the multiclass nature of the environmental movement, demands for the ownership of land and the means of production, and the democratization of the state, and for decision-making powers and the improved quality of life and living conditions for all peoples, are all linked to prior indigenous, peasant, proletarian, and urban middle-class struggles. Environmentalism does not see itself as an anticapitalist struggle in favor of "traditional" socialism but rather as a fight against the hegemonic power of the dominant economic rationality and in favor of the construction of an environmental rationality and democratization of the productive process that includes new strategies of social participation in the economic and political management of environmental resources.[39]

The environmental movement is a potential social force that can reverse the dominant economic rationality and generate the conditions for the socialization of environmental resources. At the same time, it provides political pressure that can halt destructive ecological processes while furthering environmental norms and consciousness that, in the last analysis, will enable us to appraise resources and environmental services and more closely estimate their price in terms of social costs, so that they are less vulnerable to the capitalization of nature. But the most important aspect of the environmental movement is its orientation toward the construction of a *new productive paradigm* that will establish the social and material bases for an equitable and sustainable development.

The environmental movement arose and multiplied as a result of the environmental crisis and an emerging democratic culture. Nevertheless, it is confronted with a social theory that is more oriented toward the conceptualization of the social conditions of praxis than toward strategic action for social change.[40] Thus, environmentalism has not reflected upon its own practice or its strategy for true social transformation guided by the principles and objectives of environmental rationality. Environmentalism needs a praxis that can break through the paralyzing orthodoxy and open up organizational strategies that transmit and concretize its innovating potential through the opportunities made available by a changing field of power, and thus construct a new social and productive rationality.

NOTES

1. H. M. Enzensberger, "A Critique of Political Ecology," *New Left Review*, 84, 1974, 3–31; reprinted as Chapter 1, this volume.

2. A. Granou, "Le capitalisme face à la 'non croissance,'" *Les Temps Modernes*, 236, 1973.

3. E. Leff, *Ecología y capital* (Mexico City: UNAM, 1986). A second Spanish edition (revised and extended), *Ecología y capital: Racionalidad ambiental, democracia participativa y desarrollo sustentable* (Mexico City: Siglo XXI/UNAM, 1994). The first (1986) Spanish edition was translated into English and published as *Green Production: Toward an Environmental Rationality* (New York: Guilford Press, 1995); J. Ely, "Lukács' Construction of Nature," *CNS, 1,* 1988; J. O'Connor, "Combined and Uneven Development and Ecological Crisis: A Theoretical Introduction," *Race and Class, 30*(3), 1989.

4. E. Leff, "El concepto de valor en Marx frente a la revolución científico-tecnológica," in *Teoría del valor,* ed. E. Leff (Mexico City: UNAM, 1980).

5. R. Passet, *L'economique et le vivant* (Paris: Payot, 1979); E. Leff, "Sobre la relaciones sociedad-naturaleza en el materialismo histórico," in *Biosociología y articulación de las ciencias,* ed. E. Leff (Mexico City: UNAM, 1981) (included as Chapter 1 in the 1994 edition of *Ecología y capital*).

6. E. Leff, *Green Production,* and "Ecotechnological Productivity: A Conceptual Basis for the Integrated Management of Natural Resources," *Social Science Information, 25*(3), 1986.

7. A. Schmidt, *The Concept of Nature in Marx* (London: New Left Books, 1971).

8. M. Godelier, *Economía, fetichismo y religión en las sociedades primitivas* (Mexico City: Siglo XXI Editores, 1974); C. Meillassoux, *Terrains et théories* (Paris: Editions Anthropos, 1977).

9. E. Leff, "La Dimensión cultural del manejo integrado, sustentable y sostenido de los recursos naturales," in E. Leff and J. Carabias, Coords., *Cultura y manejo sustentable de los recursos naturales* (Mexico City: CIIH-UNAM/Miguel Angel Porrúa Eds., 1993).

10. Ely, "Lukács' Construction."

11. E. Leff, "Cultura democrática, gestión ambiental y desarrollo sustentable en América Latina," *Ecología Política, 4,* 1992, 47–55.

12. D. H. Meadows et al., *The Limits to Growth* (New York: Universe Books, 1972).

13. J. O'Connor, "Capitalism, Nature, Socialism: A Theoretical Introduction," *CNS, 1,* 1988; reprinted as Chapter 9, this volume.

14. I. Sachs, *Ecodesarrollo: Desarrollo sin destrucción* (Mexico City: El Colegio de México, 1982).

15. E. Leff, *Green Production.*

16. Ibid.

17. M. O'Connor, "Codependency and Indeterminacy: A Critique of the Theory of Production," *CNS, 3,* 1989.

18. D. Faber, "Dependent Development, Disarticulated Accumulation, and Ecological Crisis in Central America," *CNS, 1,* 1988; O'Connor, "Combined and Uneven Development."

19. E. Leff, "Cultura Ecológica y Racionalidad Ambiental," in *Ecología y capital,* ed. Leff, 2nd ed., chap. 11.

20. N. Georgescu-Roegen, *The Entropy Law and the Economic Process* (Cambridge, Mass.: Harvard University Press, 1971).

21. R. A. Rappoport, "The Flow of Energy in an Agricultural Society," *Scientific American, 25,* 1971.

22. M. O'Connor, "Codependency and Indeterminacy."

23. R. Fossaert, *La Sociéte: Les Structures economiques* (Paris: Editions du Seuil, 1972).

24. L. Althusser and E. Balibar, *Reading Capital* (London/New York: Verso, 1979); L. Althusser, "Ideology and the Ideological Apparatuses of the State," in *Essays on Ideology* (London/New York: Verso, 1984).

25. Leff, *Ecología y capital.*

26. E. Morin, *La Méthode: La Nature de la nature* (Paris: Editions Anthropos, 1977) and *La Méthode: La Vie de la vie* (Paris: Editions du Seuil, 1980).

27. "As policies are available to the [economic] system, the environment gets narrower. The success of such policies will be based upon . . . the disappearance of the concept of environment, which will end up being assimilated into the system. . . . In fact, in the long run, the environment, assimilated as a permanent dimension of planning, is destined to disappear as a concrete field of action" (Sachs, *Ecodesarrollo,* 36, 53).

28. M. O'Connor, "Codependency and Indeterminacy."

29. "Human Scale Development," *Development Dialogue, 1,* CEPAUR/Dag Hammarskjold Foundation, Motala, 1989.

30. E. Leff, *Green Production,* and "Ecotechnological Productivity."

31. A. Demirovic, "Ecological Crisis and the Future of Democracy," *CNS, 2,* 1989.

32. P. Gutman, "Economía y ambiente," in *Los Problemas del conocimiento y la perspectiva ambiental del desarrollo,* ed. E. Leff (Mexico City: Siglo XXI Eds., 1986); Leff, *Ecología y capital,* and *Green Production.*

33. J. Martínez Alier, "Ecological Economics and Eco-Socialism," *CNS, 2,* 1989.

34. W. Kapp, *Social Costs, Economic Development, and Environmental Disruption* (Lanham, Md.: University Press of America, 1983).

35. Leff, *Ecología y capital.*

36. Leff, "Cultura democrática."

37. Carlos Carboni, communication to J. O'Connor, 1988.

38. J. O'Connor, "Capitalism, Nature, Socialism: A Theoretical Introduction."

39. E. Leff, "El movimiento ambientalista en México y en América Latina," *Ecología: Política/Cultura, 2*(6), November, 1988, and "The Environmental Movement and Prospects for Democracy in Latin America," in M. P. García-Guadilla and J. Blauert, eds., "Challenging Development and Democracy," special issue of the *International Journal of Sociology and Social Policy, 12*(4–7), 1992.

40. J. Ely and V. Heins, "Interview with Helmut Wiesenthal," *CNS, 3,* 1989.

Chapter 8

MARXISM
AND NATURAL LIMITS
An Ecological Critique and Reconstruction

TED BENTON

NATURAL LIMITS IN MARX AND ENGELS:
AN ECOLOGICAL CRITIQUE

I will attempt to show that certain key concepts of the economic theory of
Capital involve a series of related conflations, imprecisions, and lacunae, the
net effect of which is to render the theory incapable of adequately conceptualizing the ecological conditions and limits of human need-meeting
interactions with nature. I also will begin to indicate ways in which these
theoretical defects might be corrected. It is, perhaps, worth emphasizing
that my main aim is the constructive one of using Marx's ideas a conceptual
"raw materials" in order to move toward an ecologically adequate economic
theory. I make no claim to expository balance. I acknowledge that there is
much in the corpus of Marxian historical materialism that is readily
compatible with an ecological perspective. But my focus here is quite
deliberately upon those features of the economic theory that demand
critical transformation if they are to meet this requirement. Even here, my
view is (though I do not argue for it) that Marxian economic concepts
constitute an indispensable starting point for any theory that would adequately grasp the ecological conditions and limits of human social forms.

Notwithstanding their generally critical stance with respect to classical

157

political economy (CPE), including Ricardo's version of it, Marx and Engels sustained and deepened *those aspects* of CPE that exemplified its hostility to the idea of natural limits to capital accumulation. In the one area where Ricardo explicitly asserts a natural limit, he betrays a concession to Malthus's population theory. Marx and Engels were consistently and forcefully critical of this natural limits argument. We must conclude that in their mature economic theory, Marx and Engels were even more resistant than Ricardo to the idea of economically significant *natural* limits to capital accumulation. This is, of course, a most important claim, and it requires further substantiation. I have space here to do this only in a very truncated and schematic form.

I shall consider, first, Marx's abstract concept of the "labor process" as a transhistorical condition of human survival, and, second, his account of the forms taken by the labor process under specifically capitalist economic relations. In both cases, I shall argue, Marx underrepresents the significance of nonmanipulable natural conditions of labor processes and overrepresents the role of human intentional transformative powers vis-à-vis nature. A consequence of this is that Marx is prevented from adequately theorizing both the necessary dependence of all forms of economic life upon naturally given preconditions *and* the particularly striking and politically important form that this dependence takes with respect to specifically capitalist accumulation.

THE LABOR PROCESS

First, then, let us examine Marx's concept of the labor process. This is defined by Marx as "the everlasting Nature-imposed condition of human existence." At this level of abstraction, Marx excludes consideration of the historically variable social relations between persons in the process, whether these be relations materially required by it, or relations (such as those arising out of property ownership) socially imposed upon it. Thus defined, the "elementary factors" of the labor process are: (1) the personal activity of man, that is, work itself; (2) the subject of that work; and (3) its instruments.[1]

The process itself is an activity in which these elements are brought into appropriate relationships to one another and set in motion: "In the labour-process . . . man's activity, with the help of the instruments of labour, effects an alteration, designed from the commencement, in the material worked upon. The process disappears in the product; the latter is a use-value, Nature's material adapted by a change of form to the wants of man."[2] Marx here attempts to characterize human need-meeting activity upon nature as a process in which human activity employs instruments in order to bring about a change in some material object or substance. The change so wrought is intended to fit the object or substance to function as a means to satisfy some human need or want. Marx concedes that some very elementary transactions with nature do not require artificial implements, and here human limbs themselves can be regarded as playing the part of "instruments of production." The "subject" of labor—the thing or material worked upon—may be "spontaneously provided by nature," or, more commonly, it will have been "filtered through the past labour," in which case Marx speaks (somewhat misleadingly) of "raw material." Raw materials are divided into two categories—those that form the "principal substance" of the product, and those (such as fuels, dyestuffs, etc.) that enter into the labor process, but do not go to form the "principal substance" of the product. Marx calls these "accessory" raw materials. Henceforth, I shall speak of "raw materials A" and "raw materials B," respectively.

So far, as we have noted, Marx recognizes among the instruments of labor both human limbs themselves, and objects and materials "which the labourer interposes between himself and the subject of his labour, and which serve as the conductor of his activity."[3] But what of other elements and features of the environment within which the labor process takes place? Marx recognizes that the earth itself (which "furnishes a locus standi to the labourer and a field of employment for his activity"[4]) as well as results of previous labor such as workshops, canals, roads, and so forth may be conditions for labor processes, without entering directly into them. Marx proposes to include these among the instruments of labor ("in a wider sense"). To recognize, as Marx himself does, that these "instruments of labour" in the wider sense may be naturally given or the results of previous labor, I shall speak of "natural" and "produced" conditions of the labor process, respectively.

Central to Marx's abstract concept of the labor process, therefore, is the notion of a *raw material A* undergoing a *transformation* to yield a *use value*. This transformation is the outcome of a *human labor* that involves the utilization of *raw material B* and *instruments of labor* to achieve its purpose. The process involves both human intentional activity, and a range of distinct materials, substances, and other nonhuman beings and conditions. There is, then, a primary, threefold classification of the elements of the labor process, in which the nonhuman elements are resolved into the two categories of "subject" of labor (i.e., that which is worked upon) and instruments of labor (the conductors of intentional activity). There is also a secondary classification in which these major categories are further subdivided (raw materials A and B, natural and produced conditions, etc.). What is absolutely clear, however, is that both primary and secondary classifications allocate elements in the labor process to conceptual categories on the basis not of material characteristics but of their relationship to the purpose of the labor process itself. One and the same item may, at different times, figure as product, instrument, and raw material of different labor processes. Which category it falls into on any particular occasion will be a function of what I shall call the *intentional structure* of the labor process.

Now, it is immediately clear that the intentional structure of the labor process, is for Marx, a transformative one. It is plausible to suppose that Marx's model is handicraft production of some kind. Carpentry, for example, could be readily represented as having just such an instrumental–transformative intentional structure. With some modification, the representation might do for productive labor process in general, though with the important reservation that in industrial labor processes the intentionality that assigns each element to its place in the structure is not that of the individual agents in the process. However, and this is the key point, Marx's conceptualization is supposed to represent not just *one broad type* of human need-meeting interaction with nature, but, rather, a universal "nature-imposed condition of human existence." Marx does, indeed, recognize such activities as felling timber, catching fish, extracting ore, and agriculture as labor processes. But he constructs his general concept of the labor process as if these diverse forms of human activity in relation to nature could be assimilated to it.

Ecoregulation

Let us first consider agricultural labor processes. My account, here, assumes such labor processes taking place on land already cleared and prepared, and using seed or stock animals that already embody past labors of breeding and selection. (To work without these assumptions is, in fact, less favorable to Marx.) My argument, at this stage in the argument, also abstracts both from the property relations (or "social relations of production") within which agricultural labor processes might be formed and from the great range of specific agricultural technologies that have been and might yet be realized. I am, of course, very far from thinking that such considerations are unimportant. On the contrary, working at this level of abstraction will, I hope, help to illuminate the precise importance of such considerations (e.g., in relation to the ecological consequences of "green revolution" agricultural technologies and the modern capitalist "industrialization" of agriculture).

In agricultural labor processes, by contrast with productive, transformative ones, human labor is not deployed to bring about an intended transformation in a raw material. It is, rather, primarily deployed to sustain or regulate the environmental conditions under which seed or stock animals grow and develop. There *is* a transformative moment in these labor processes, but the transformations are brought about by naturally given organic mechanism, not by the application of human labor.[5] Agriculture, and other "ecoregulatory" labor processes thus share an intentional structure that is quite different from that of productive, transformative labor processes. This is so even where, as in modern capitalist agriculture, the forms of calculation employed by economic agents, and the economic dynamic of the process, more closely resemble the productive–transformative model. Indeed, the primary source of the ecological problems of modern capitalist agriculture lies precisely in the tension between these features and the constraints imposed by its intentional structure as what I call an "ecoregulatory" practice. The extent of the difference between ecoregulatory and productive practices is the measure of the inadequacy of Marx's abstract concept of the labor process, which, as we have seen, assimilates all labor processes to a "productive" model. For my purposes here the key distinctive features of ecoregulatory practices are as follows:

1. Labor is applied primarily to optimizing the *conditions for* transformations, which are themselves organic processes, relatively impervious to intentional modification. The "subject of labor" (in Marx's terminology) is therefore *not* the raw material that will become the "principal substance" of the "product" but rather the conditions within which it grows and develops.

2. This labor, optimizing the conditions for organic growth and development, is primarily (once agriculture is established) a labor of sustaining, regulating, and reproducing, rather than of transforming (e.g., maintaining the physical structure of the soil as a growing medium, maintaining and regulating the supply of water, supplying nutrients in appropriate quantities and at appropriate times, reducing or eliminating competition and predations from other organic species, etc.).

3. The spatial and temporal distributions of laboring activity are to a high degree shaped by the contextual conditions of the labor process and by the rhythms of organic developmental processes.

4. Nature-given conditions (water supply, climatic conditions, etc.) figure both as *conditions* of the labor process, *and* as *subjects* of labor, yielding a catergory of "elements" of the labor process not readily assimilable to Marx's tripartite classification (labor, instruments of labor, raw materials).

These four features draw attention to the extent of the *dependence* of ecoregulatory practices upon characteristics of their contextual conditions, which are the principal "subjects" of their labor, and on the organic processes they aim to foster. For any specific technical organization of agriculture, these elements in the process are relatively impervious to intentional manipulation, and in some respects they are absolutely nonmanipulable. For example, the incidence of radiant energy from the sun is absolutely nonmanipulable. Labor processes in agriculture are therefore confined to optimizing the efficiency of its "capture" by photosynthesizing crop plants, or complementing it with artificial energy sources. In this case, nonmanipulability is a consequence of the scale of the natural mechanisms involved, compared with the causal mechanisms that can be mobilized by human interventions. In other cases, such as climatic conditions, human interventions may, indeed, be cumulatively large enough in scale to have *effects* (e.g., the notorious greenhouse effect), but these effects do not and arguably cannot amount to intentional manipulation. The combination of epistemic obstacles and the problem of scale confine us to having effects on weather systems that are predominantly unintended and largely unwanted.

In yet another class of cases—organic processes of growth and development themselves—it might be argued that recent and foreseeable developments in biotechnology are in train to eliminate what I have identified as key distinctive features of ecoregulatory practices. The widespread artificial use of hormones to intervene in organic developmental processes and genetic engineering technologies both seem to point in this direction. My response, here, would be to suggest that the newer biological technologies have been "sold" within a voluntaristic-Promethean discourse that has invariably occluded or rendered marginal the limits, constraints, and unintended consequences of their deployment in agricultural systems. It is, for example, widely recognized among geneticists that a genetic modification that enhances the utility of an organism for agricultural purposes is generally accompanied by countervailing "costs": higher yield as against lower resistance to disease, or greater vulnerability to enironmental stresses, for example. Organisms are not mere aggregate expressions of contingently connected and freely manipulable genetic particles.

Let us now turn briefly to the intentional structures of such primary labor processes as hunting, gathering, mining, and so on. These cases are more like production than ecoregulation in one respect: labor is applied directly to the object or material that is the intended repository of use value. However, there are two major differences between these labor processes and productive ones. First, the conversion of the "subject of labor" into a use value cannot be adequately described as "Nature's material adapted by a *change of form* to the wants of man." This conversion is rather a matter of selecting, extracting, and relocating elements of the natural environment so as to put them at the disposal of other practices (of production or consumption). These primary labor processes, then, *appropriate* but do not transform. Second, they, like ecoregulatory processes, are highly dependent on both naturally given contextual conditions and the properties of the subjects of labor. In these practices, the place of principal and accessory raw materials is taken by "naturally given" materials or beings, whose location and availability are relatively or absolutely impervious to intentional manipulation.

Ecoregulation and primary appropriation, then, have intentional structures that cannot adequately be characterized in the terms of Marx's abstract conceptualization of the labor process. This is because of Marx's implicit "Procrustean" overgeneralization of the intentional structure appropriate to productive, transformative labor processes. Labor processes—whose intentional structure emphasizes the dependence of labor on non-manipulable conditions and subjects—in which labor adapts to its conditions, sustains, regulates, or appropriates its subjects, as distinct from transforming them, are given *no independent conceptual specification*. The significance of their place in the overall "metabolism" between human populations and their natural conditions becomes literally unthinkable.

Capitalist Production

Of course, Marx's abstract concept of the labor process as a transhistorical condition of human existence is not central to his concerns in *Capital*. It could very plausibly be argued that I have made too much of his problems with this one concept. Marx and Engels themselves tended to work with a view of the historical development of the "forces of production" in which the nontransformative labor processes I have just discussed are seen less as "moments" in the total economic life of any particular society, than as forms of interaction with nature that characterized whole previous epochs of human history. Simple appropriation or collection is characteristic of human societies at the lowest level of development of their productive powers, with agriculture marking the historical acquisition of powers to wrest from nature means of subsistence that it would not otherwise have provided. This, in turn, is a condition for a further division of labor and the eventual emergence of modern industrial production.

Marx and Engels, like most of their contemporary economic and social theorists, were profoundly impressed by the transformative power of modern industrial production. It was understandable that they should focus upon it and the social relations implicated in it as the central historical dynamic of their time. However, what Marx and Engels never adequately theorized (without, on the other hand, ever quite forgetting) was the extent to which this massive and dynamic sector of nineteenth-century European economic life remained tied to ecoregulatory and primary-appropriative labors as the necessary sources of energy, raw materials, and food, and so, also, to a range of nonmanipulable contextual conditions.

To show how and why this was so, it will be necessary to consider the way in which Marx conceptualized specifically *capitalist* production. Central to his analysis is the distinction between capitalist production considered as a process of producing use values and as a process of producing exchange values. In its former aspect, capitalist production is a labor process in which a specific kind of useful labor is set to work "concretely" to transform a raw material into a specific useful product. In its latter aspect, capitalist production is a process of exploitation of "abstract" labor that aims at a quantitative increase in exchange value. The second intentional structure is considered to be abstract in the sense that the material properties of the product, the character of the labor that shapes it, and the nature of the want it satisfies are all quite irrelevant to the central purpose of a purely quantitative increment in the value of the product. The social–structural conditions governing economic action under capitalism require that this second, value-maximizing intentional structure must be superimposed upon, and predominate over, the intentional structure of production in its aspect as a utility-producing labor process. This asymmetrical relation between the two structures is the source of the liability of capitalist economic relations to generate severe dislocations between what is produced and what is needed, and a whole range of irrationalities in the global allocation of labor and material resources in relation to human needs.[6] Although a capitalist economy must in the long run balance production and consumption across the various sectors of production and meet the minimal subsistence needs of its laboring population, Marx's analysis is an attempt to show why this must be an unstable, disruptive, and crisis-ridden process, rather than a matter of harmonious regulation by a beneficent "invisible hand."

Probably because Marx was convinced that the overall dynamics of capital accumulation and the contradictions of this process were rooted in the abstract, value-creating aspect of capitalist production, it was this aspect that constituted the overwhelming topic of analysis in *Capital*. Marx showed relatively much less interest in labor processes as "concrete" combinations of specific kinds of labor with specific instruments. Certainly he did give an important place to the distinctively capitalist tendency to replace living labor with machinery, and to transform the technical basis of the labor process. But his crucial interest was in the consequences of these tendencies for capital accumulation considered in value terms, and for the development of the antagonism between capital and labor. This focus has been continued by Marx's successors in the "labor-process debate," in ways which have shifted attention away from labor processes—*including* those involved in capitalist production—as social forms of interchange with nature.[7]

The Concrete Intentional Structure

If we turn, now, to consider capitalist production from the standpoint of its *subordinate*, yet necessarily present intentional structure as concrete labor process, it is possible to show that Marx's account is defective *even with respect to productive/transformative* labor processes. There are five main features of productive labor processes that tend to be either under-theorized by Marx or simply left out of account. These are:

1.   The material nature of both instruments of labor and raw materials will set limits to their utilization/transformability in line with human intentions. Recognition of these limits is a condition of effective practice.

2.   Although the immediate source of raw materials and instruments of labor may be earlier labor processes, indirectly all must have their source in appropriation from nature (in some form of "collection"). Persistence of specific production processes is therefore dependent not only upon ancillary production processes, but also upon appropriation from nature.

3.   Labor itself is an indispensable element in the labor process. Marx treats the activity of laboring as the "consumption" of the laborer's capacity for work, or "labour-power." This, in turn, is treated as the product of the prior labor of producing the laborer's means of subsistence. Marx tended, like the political economists, to treat the production and reproduction of workers themselves as equivalent to the (capitalist) production of their means of consumption. This excludes from recognition household work of reproduction, nurturing, and so on, and has rightly been criticized by feminists. Inconclusive as it has been, the "domestic labor debate" has at least demonstrated that distinctive labor processes, conducted *outside* the sphere of capitalist economic relations as they are characterized in *Capital,* are presupposed in the appearance and reappearance of workers with labor power for sale in capitalist labor markets. Presupposed by the productive labor processes of capitalism are ("domestic") labor processes that have a quite different intentional structure.[8] In their dependence on organic processes of reproduction and development they resemble ecoregulatory practices, but in their affective and normative content they are quite unlike all other labor processes. Marx's implicit assimilation of these processes to productive processes complements his anti-Malthusian and more general reluctance to recognize "natural limits."

4.   Although the dependence of productive practices upon contextual conditions is less apt to figure in the calculations of agents, these practices, no less than primary appropriation, ecoregulation, and human reproduction, *are* dependent upon such conditions. Marx recognizes that these conditions include some that are naturally given and others, such as factories and roads, that are products of past labor. But in recognizing the necessity of these conditions, Marx simultaneously fails to recognize their significance by including them *within* the category of "instruments of production." These conditions cannot plausibly be considered "conductors" of the activity of the laborer. By definition, they do not *enter into* the labor process at all. Moreover, the subjection to human *intentionality* that

is implicit in the concept of an "instrument" is precisely what *cannot* be plausibly attributed to these contextual conditions of production. This is particularly true of naturally given geological, geographical, and climatic conditions. Even in the case of productive/transformative labor processes, then, the conceptual assimilation of contextual *conditions* of the labor process to the category "instruments of production" has the effect of occluding the essential dependence of all labor processes upon at least some nonmanipulable contextual conditions. Marx, like Ricardo before him, was able to get away with his occlusion to the extent that such contextual conditions of production could be taken as unproblematically "given," either being insusceptible to or simply not requiring intentional manipulation under conditions then prevailing.

5. Finally, Marx's intentional, or "functional," classification identifies the elements in the labor process in terms of those properties (causal powers, liabilities, and tendencies) in virtue of which they are acted upon or utilized by human agents in order to achieve their purposes. In any actual labor process these properties of its elements will be only a *limited subset* of the properties really possessed. The remaining properties constitute an indefinitely large "residual" category which, from the standpoint of the calculations of the agents involved in the labor process, may be known or unknown, relevant or irrelevant to the achievement of the immediate purposes of the labor process. Insofar as this class of properties did not figure in Marx's characterization of the labor process, he shared the blindness of agents themselves to the sources of naturally mediated unintended and unforeseen consequences of specific practices of activity upon nature. This point will be developed in what follows.

These five general conditions all point in the same direction. In each case, Marx's conception even of productive labor processes is shown to exaggerate their potentially transformative character, while undertheorizing or occluding the various respects in which they are subject to naturally given and/or relatively nonmanipulable conditions and limits. From the standpoint of an ecological critique, three important corrections thus need to be made to Marx's conceptualization. First, contextual conditions should be conceived separately from the instruments of labor, as an independent class of "initial conditions." Second, the continuing pertinence of these contextual conditions to the *sustainability* of production needs to be incorporated, as with ecoregulatory practices. This is significant in that it renders thinkable the possibility that these conditions might cease to be spontaneously satisfied, and so require the ancillary labor process of restoring or maintaining the environmental conditions for productive sustainability. André Gorz has provided an interesting theorization of this possibility in the shape of a postulated environmental foundation for a tendential fall in the rate of profit.[9] Third, some of the naturally mediated unintended consequences of the operation of labor processes may impinge upon the persistence or reproduction of its contextual conditions and/or raw materials. Where raw materials are absolutely limited in supply, or can be replenished only at a definite maximum rate (e.g., because of their dependence upon organic developmental processes), the labor processes may undermine their conditions of sustainability in ways suggested under Point 2 above.

Where, on the other hand, causal mechanisms are set in motion by the labor process that are *extrinsic to the achievement of its purposes,* future sustainability may be undermined by another route. Energy and materials entering into the productive process but not embodied in the product, for example, are conceptualized by Marx as "necessary" raw materials. His account of the intentional structure of production ignores the "further adventures" of these materials or their residues once they have played their part in the achievement of the intrinsic purpose of the labor process. This abstraction on Marx's part is justified to the extent that these "further adventures" bear no relation to the sustainability of the labor process in question. However, once the dependence of productive labor processes on contextual conditions is explicitly recognized, then the possibility that they may be undermined by their own naturally mediated unintended consequences is open to investigation. Among these unintended consequences may be the effects of accessory raw materials and their residues as well as unutilized energy releases upon water supplies, atmospheric conditions, climatic variables, and so on.

So far, I have tried to demonstrate that in a number of respects Marx's account of capitalist production employs a limited and defective concept of productive labor processes. Each of these limits and defects contributes to an overall exaggeration of the potential transformative power of such labor processes, at the expense of any full recognition of their continued dependence upon and limitation by other nonproductive labor processes,[10] by relatively or absolutely nonmanipulable contextual condition, and by naturally mediated unintended consequences. I have not, so far, considered the implications for my argument of the *dual* intentional structure of capitalist production: of its status as a process of self-expansion of value. I will reserve this consideration for a later stage in the argument.

SOURCES OF THE "PRODUCTIVIST" IDEOLOGY

Nature, Capitalism, and Emancipation

Next, I shall try to show that these weaknesses in Marx's account of the labor process are not merely contingent "errors," particular failings of insight. On the contrary, they are coherent with significant strands in the wider theoretical perspectives of Marx and Engels, and they have a certain plausibility given the historical location of their thinking. First, Marx and Engels were disposed, especially through their critiques of Malthus, to reject as necessarily conservative "natural limits" arguments. While they were firmly committed to the view that capital accumulation *was* subject to outer limits, these limits were theorized as *internally* generated by the contradictory social–relational structures of capitalist economies, and mediated through class struggles.

Second, notwithstanding their systematic moral critique of capitalism, and their analysis of its transitory nature, Marx and Engels also held an "optimistic" view of its historical role as preparing the conditions for future human emancipation.[11] In its *progressive* historical role, capitalism accelerates the development of the forces of production to the point where transition to a realm of freedom and abundance becomes a real historical possibility. "Development of the productive forces of social labour is the historical task and justification of capital. This is the way it unconsciously creates the requirements of a higher mode of production."[12] Modern industrial production, fostered by capitalist economic relations, is a precondition for the future communist society. The "historical task" of capitalism is precisely to transcend the conditional and limited character of earlier forms of interaction with nature. Emphasis upon the transformative powers of human social labor, as embodied in industrial capitalist labor processes, is, then, an intrinsic element in Marx's overall view of the historical process. Moreover, as the above quotation also suggests, this is something that bears not only upon the "historical task" of capitalism, but on the very conceptualization of the postcapitalist future itself.

Marx and Engels were famously reticent about the character of this communist future. But the brief remarks they did allow themselves always gave central place to the emancipatory potential of a communal appropriation of nature that presupposes the inheritance of highly developed productive forces from capitalist prehistory. Engels's remarks in *Socialism: Utopian and Scientific* are a striking example:

> The whole sphere of the conditions of life which environ man, and which have hitherto ruled man, now comes under the dominion and control of man, who for the first time becomes the real, conscious lord of Nature, because he has now become master of his own social organization. The laws of his own social action, hitherto standing face to face with man as laws of Nature foreign to, and dominating him, will then be used with full understanding, and so mastered by him.[13]

The view of emancipation implicit in this passage and others like it is roughly this: In earlier stages of history, humans have suffered a doubly conditioned lack of autonomy. Insofar as their transformative powers vis-à-vis nature have been limited in their development, they have been at the mercy of, dominated by, the forces of external nature. But superimposed upon this source of domination has been another, rooted in society itself, experienced as a "second nature." With the historical development of human social powers vis-à-vis nature there arises the possibility that the tables can be turned with respect to both sources of oppression: humans can acquire communal control over their own social life, and through that, over nature itself.

But if the acquisition of human anatomy presupposes control over nature, this suggests an underlying antagonism between human purposes and nature: either we control nature, or it controls us! There is no room, apparently, for symbiosis, peaceful coexistence, mutual indifference, or other imaginable metaphors for this relationship. At first encounter, it might seem that this view of an underlying antagonism in our relation to nature does, after all, embody a notion of naturally imposed limits. But this is a misleading impression. For Engels, in this passage at least, progress in our productive powers is achieved to the extent that we incorporate what was previously encountered as an external limit within the sphere of conscious human control. Sometimes, as in this passage, Marx and Engels speak as if this process could be extrapolated without limit. The metaphor in the early *Manuscripts* of a "humanization of nature" seems to suggest a potentially residue-less subjection of the natural world to human intentionality. Elsewhere there is a recognition that *some* element of "struggle" with nature for the necessaries of life is inevitable, the content of emancipation being given in the reduction to a minimum of the time taken up in this struggle. Either way, the possibility of human emancipation is premised upon the potential for the transformative, productive powers of associated human beings to transcend apparent natural limits, and to widen the field of play for human intentionality. The coherence between this notion of emancipation, the terms in which Marx and Engels criticize Malthus, and the defects in Marx's concept of the labor process should now be apparent.

Spontaneous Ideologies of Industrial Capitalism

I shall shortly return to this question, but first I want to consider a further way of making intelligible Marx and Engels's systematic exaggeration of the potential transformative power of human action in relation to nature. This approach to the problem makes use of the idea, often employed by Marx and Engels themselves, that certain structures of interaction present to actors who participate in them forms of appearance that are systematically misleading. Actors affected by such forms of appearance will tend to hold mistaken or distorted beliefs about their own activities. Such patterns of mistaken or distorted belief we may call "spontaneous ideologies." Insofar as participants in productive labor processes, especially industrial ones, are not required to attend to the maintenance or restoration of contextual conditions, are not *in fact* confronted with absolute shortages of raw materials, and can ignore the extrinsic, unintended consequences of their practices, then to this extent they are liable to exaggerate their potential transformative powers. Marx could in this way be understood as a victim of a widespread spontaneous ideology of nineteenth-century industrialism.

But this is by no means the whole story. Here we must return to a consideration of the overriding intentional structure of capitalist production: the intentionality of self-expanding value. As we have seen, Marx and Engels saw the development of the forces of production under capitalism as itself only a secondary—though still necessary—byproduct of this primary intentional structure. Now, the very feature that defines this principal dynamic of capitalist accumulation is indifference to the qualitative character of the labor process. Production on an ever-growing scale is a requirement for the survival of individual capitals, and of the system as a whole. But what is produced, and with what resources are entirely secondary to the purpose of quantitative maximization of exchange values. The labor theory of value, which Marx adopted, albeit with important modifications, from his predecessors in classical political economy, is the central conceptual device through which the limits, contradictions, and crises of capital accumulation are rendered thoroughly social–relational. For Marx, as for Ricardo, the labor theory of value either excludes natural scarcity from consideration, or allows it to be recognized only in the form of its displaced manifestation within the internal, social–relational structure of the economy.

It is tempting to see Marx's focus on capitalist production as a process of maximization of exchange value by means of the exploitation of labor, his focus on the social relations of production at the expense of the labor process as yet one more element in his flight from any recognition of "natural limits." On this reading, the spontaneous ideology of nineteenth-century industrialism is overdetermined by a spontaneous ideology of capitalism as a (naturally) limitless process of self-expansion of value. The blindness to natural limits already present in the industrial ideology is compounded and intensified by the overriding intentional structure, with its indifference to the concrete character of raw materials, labor, *or* product.

TOWARD A GREEN HISTORICAL MATERIALISM

But now we have reached the point in the argument where we can begin to see the potential explanatory *fruitfulness* of Marx's critical account of capitalist accumulation, once defects in his concept of the labor process are corrected. Only if this part of the argument works can the effort of a critical encounter be shown to pay off!

Relativizing the Nature–Society Connection

The first step in this more positive task of theoretical reconstruction is to call into question the terms of Marx's and Engels's responses to Malthus. More broadly, it is necessary to recognize that "natural limits" epistemic conservatisms can be countered effectively without wholesale retreat into social constructionism. In effect, this is the error Marx and Engels make. For them, "relative surplus population," or "the reserve army of labour," is a *consequence* of the dynamic tendency of capital accumulation. The form taken by the argument in Marx's *Capital* makes the surplus population an effect of the tendency of constant capital to rise as a proportion of total capital. But, in fact, this only follows on the basis of implicit assumptions about the rate at which the working population reproduces biologically. Marx does not acknowledge these assumptions.

However, at least one element in Marx's and Engels's argumentative strategy against Malthus can be endorsed. This is their commitment to relativizing Malthus's law to specific historical epochs (or forms of society). As I have shown,[14] in effect Marx and Engels take historical/social relativization to imply some form of social constructionism. This is not, I believe, required. What *is* required is the recognition that each form of social/economic life has its own specific mode and dynamic of interrelation with its own specific contextual conditions, resource materials, energy sources, and naturally mediated unintended consequences (forms of "waste," "pollution," etc.). The ecological problems of any form of social and economic life would have to be theorized as the outcome of this specific structure of natural/social articulation.

This approach avoids both the Scylla of epistemic conservatism and the Charybdis of "social–constructionist" utopianism. Each form of social and economic life is understood in terms of its own specific contextual conditions and limits. These conditions and limits have real causal importance in enabling a range of social practices and human purposes that could otherwise not occur, and also in setting boundaries and limits to their sustainability. At the same time, giving full theoretical recognition to contextual (including natural) conditions and limits in this way opens up alternative ways of conceptualizing the relationship between emancipatory strategies and natural limits.

First, and most important, we can look again at the crucial assumption underlying the technological optimism Marx and Engels shared with many other theorists of development (ideas that are still very influential). This is the assumption of an intrinsic antagonism between the fulfillment of human purposes, on the one hand, and the forces of nature, on the other. Although there were, as we shall see, moments when Marx and Engels criticized this idea, it remains as a presupposition of their view of the relationship between the historical development of the forces of production and the ultimate achievement of human emancipation. Natural conditions and limits tend to be regarded as a primary source of human heteronomy, the progressive function of the development of the forces of production consisting in their *transcendence* of limits by incorporating natural conditions within the sphere of human intentionality: a domination or control of nature.

By contrast, the reconceptualizations of labor processes that I am advocating permit an explicit recognition of the ways in which naturally given processes, mechanisms, and conditions *make possible* human need-meeting practices that otherwise could not occur. But in any realist or materialist approach, enabling conditions must be understood as simply the obverse side of the coin from limits or constraints. A power conferred on human agents by a specific social relation to a natural condition or mechanism will also be bounded in its scope by that self-same relation. If, for example, a naturally given water supply, in the form of a river, is utilized by a human population both for agricultural irrigation and fishing, it figures as an enabling condition for both practices. Insofar as human needs are met and purposes fulfilled by both practices, the combination of the socially established technology with the naturally given condition can be seen as emancipatory. However, once this pattern of interaction with nature is established its continuation is subject to definite limiting conditions. High levels of fertilizer runoff, or irregularities in the outflow of water from irrigation dikes, for example, will have their effects on fish populations in the river. Only a reconceptualization of labor processes along the lines I am suggesting can render analyzable such complex patterns of enablement and constraint that are built into all forms of human interaction with nature. If follows from considerations such as this that "natural limits" arguments are not in conflict with emancipatory projects as such. Provided "natural limits" are conceptualized in ways that recognize their historical, geographical, and social relativity, they are compatible only with *utopian* would-be emancipatory strategies.

Second, since natural limits are themselves theorized, in this approach, as a function of the articulated combination of specific social practices and specific complexes of natural conditions, resources, and mechanisms, what constitutes a genuine natural limit for one such form of nature–society articulation may *not* constitute a limit for another.[15] In other words, a fundamental reorganization of the form and dynamics of the interrelation of a society with external nature may have the effect of transcending what were, given the previous mode of appropriating nature, real natural limits. Clearly, if we think in terms, for example, of nonrenewable resources as a natural limit, a society that shifts its resource base, or builds resource recycling into the intentional structure of its labor processes, may effectively transcend what previously were encountered as limits. Again, to recognize that specific social and economic forms of life encounter real natural limits is to concede nothing to natural limits conservatism. On the contrary, it may provide the beginnings of a powerful argument for *transforming* the prevailing pattern of nature–society interaction. Of course, the new form will also be a specific combination of enabling conditions and constraints (with respect to whatever human social purposes prevail within it). Different technical bases for society can be understood in this way as delimiting specific alternative patterns of possibility for further human development. Theorizing such alternative possibility spaces can help in thinking about "development" not as a unilinear process of quantitative expansion of the forces of production, but rather in terms of a range of qualitatively different ways of realizing human social possibilities.

Finally, there is a third respect in which the approach I am advocating illuminates the relation between emancipatory strategies and natural limits. By giving explicit theoretical recognition to relatively or absolutely nonmanipulable conditions and elements in labor processes, one throws into relief a distinction between technologies that enable a *transcendence* of of naturally imposed limits and technologies that enhance *adaptability* in the face of natural conditions impervious to intentional action. Adaptive, as distinct from transformative, technologies are at work in some of the most fundamental and distinctive features of human ecology: the building of shelters, clothing, the use of artificial means of transport, and so on can be seen as sociocultural extensions of biological features such as warm-bloodedness that facilitate both survival and well-being in the face of a spectacular range of environmental conditions. A strategic focus on adapt-ability-enhancing technologies may be no less emancipatory, and is cer-tainly likely to be far more sustainable, than the transformative focus that predominates in our civilization.

Rethinking Labor Processes and Ecological Politics

Two proposals for the positive work of theoretical reconstruction emerge from these admittedly extremely abstract and entirely provisional consid-eration. The first would be an extension of what I have tried to do above by way of providing a provisional typology of labor processes and repre-senting their intentional structures. This would give special attention to the character of labor processes as modes of social appropriation of nature, so displacing the centrality of the focus on social–relational aspects in the labor process debate as currently conducted. To do this is not in any way to argue for "demotion" of social–relational issues; the point is rather to suggest that the patterning and dynamics of power relations and social conflicts in the labor process, and in the wider society, will be viewed very differently on the basis of a reconceptualization of the labor process itself.

Two examples of this spring to mind immediately. One is that the structure of possibilities open to workers in resisting new technologies is highly dependent upon material properties and causal powers of technologies, many of which are *extrinsic* to the intentional structure of the labor process (i.e., properties and powers other than those in virtue of which the technology is employed). A second example concerns the wider enironmental unintended consequences of labor processes. Some of these have highly differentiated impacts on the quality of life for different categories of social actors. The political significance of this in terms of the patterning of conflict along lines of cleavage related to gender, residential location, occupational situation, life-style, and so on needs to be understood, and integrated with the more traditional focus on lines of class cleavage arising "at the point of production." For example, "green revolution" technologies in agriculture, depending on the political, social, and economic conditions of their introduction, are liable to produce direct consequences for distributional inequality and rural class structure. Owing to the relatively high levels of capital outlay required, economies of scale, and differential access to credit, larger farming units typically benefit more, and more quickly, than smaller ones. The tendency toward capitalist relations in agriculture is accelerated, peasant farmers are dispossessed and converted into landless rural or urban laborers, and so on. These processes are well recognized and much discussed. But they exist alongside and articulate with the socioeconomic effects of *naturally mediated* unintended consequences of the new agricultural technologies. Again, there will be much variation from locality to locality, but, typically, the greater intensity of farming and associated simplification of local ecosystems will put increasing pressure on traditional practices that rely on various types of primary appropriation from natural and seminatural habitats (the collection of firewood, fishing, hunting, gathering of fruits and seeds, etc.). Pesticide and nutrient seepage into water supplies and fisheries, too, may contribute to these processes, and impose other costs on the quality of life of agricultural and nonagricultural populations alike. Again, pesticide residues and chemical additives associated with forms of food processing and storage that are themselves adjuncts of "green revolution" farming techniques have implications for the diet and and health of consumers of agricultural goods well beyond the boundaries of those social groups affected by the immediate socioeconomic consequences of agricultural reorganization. It remains to be seen what forms of oppositional alliance could be established among such disparate groupings and what goals they might formulated, but the necessary analysis cannot even begin until the category of "naturally mediated unintended consequences" is fully integrated into social and economic theory.

A second possible avenue for further theoretical work would be to reconceptualize the Marxian typology of modes of production as articulated combinations of forces and relations of production. This would entail, *in each case*, not only specifying the social–relational aspects of each mode and the intentional structure of the labor process (in Marx's terms, the characteristic forces and relations of each mode), but also *complementing* this with a nonintentional characterization of the contextual sustaining conditions and liability to generate naturally mediated unintended consequences.[16] Each mode would, in other words, be thought of as instantiating the specific form of nature–society interaction, and as having its own distinctive ecological "niche." Each mode must be conceptualized in terms of its own peculiar limits and boundaries, and its own associated liabilities to generate environmental crises and environmentally related patterns of social conflict.

Such a project would help to do away with two highly pervasive misconceptions of our contemporary environmental crisis. First, as this line of analysis shows, the contemporary crisis cannot be understood as the direct, unmediated consequence of either "population" or "industrialization." Environmental impacts are a function of complex combinations of social practices and their contextual conditions, not of persons and their appetites. Second, we are confronted with at least two broad categories of social–structural embodiment of industrial societies: "Western Capitalist" and "State Socialist," together with several variant forms that do not fit easily into either category. It is clear that *each* of these historical forms has its environmental contradictions, and equally clear that their patterns, dynamics, and emergent lines of social and political cleavage are very different. It should also be noted that there is widespread evidence of often catastrophic environmental contradictions affecting both preindustrial and nonindustrial economic formations. As the foregoing points imply, it is a mistake to suppose that capitalism is the root of *all* ecological evil. I think it can be shown that capitalism is a mode peculiarly liable to ecological crisis, but it must not be forgotten that other modes, too, have their own distinctive ecological crisis tendencies.

Building on Marx and Engels

Finally, I shall return, briefly, to exegesis of Marx and Engels on these questions. A "productivist," "Promethean" view of history is widely attributed to them. I have, for the purposes of the above argument, tended to accept this reading, and have delved into some of the conceptual "substructure" that underpins it. But while I remain committed to the above ecological critique of their work, it is also necessary to recognize the basis for another, quite different reading, in which their explicit or tacit acknowledgment of some of these ecological arguments may be emphasized. Engels, for example, in his critique of social Darwinism, adopts a much wider concept of "production" than the "transformationist" one I have criticized.[17] Here, he contrasts collection with production, conceiving of the latter as any practice by which human beings prepare means of life that nature itself would not have provided. This does not necessarily involve the "raw materials/instrument/transformation" pattern of production in the narrow sense and is consistent with adaptive and regulative models of natural–social interactions as well as with domination/control/transformation models.

Both Engels and Marx do also have ways of characterizing the future society that deliberately avoid the "triumphalism" and utopianism of the "productivist" account. One very striking example is also provided by Engels:

> Thus at every step we are reminded that we by no means rule over nature like a conqueror over foreign people, like someone standing outside nature—but that we, with flesh, blood and brain, belong to nature, and exist in its midst, and that all our mastering of it consists in the fact that we have the advantage over all other beings of being able to know and correctly apply its laws.[18]

The following well-known passage from *Capital*, volume 3 is easily read as confirming the Promethean view of a historical struggle to subdue and control the forces of nature:

> With his development this realm of physical necessity expands as a result of his wants; but, at the same time, the forces of production which satisfy these wants also increase. Freedom in this field can only consist in socialised man, the associated producers, rationally regulating their interchange with nature, bringing it under their common control, instead of being ruled by it as by a blind power.[19]

But if we read this passage as postulating not the bringing of *nature* under common control, but rather, "*interchange* with nature," then it is quite consistent with the idea of a form of interaction with nature that integrates ecological self-regulation within its intentional structure. There is, indeed, other textual evidence to support such a reading. An important example, to which I shall return, is Marx's discussion of the deleterious effects of capitalist agriculture on soil fertility: "By this action it destroys at the same time the health of the town labourer and the intellectual life of the rural labourer. But while upsetting the naturally grown conditions for the maintenance of that circulation of matter, it *imperiously calls for its restoration as a system, as a regulating law of social production,* and under a from appropriate to the full development of the human race."[20]

These passages suggest that both Marx and Engels did recognize the transhistorical necessity of human dependence upon naturally given conditions and limits to their social activity. Crucially, the latter quotation from Marx quite explicitly advocates ecological sustainability as a "regulating law" that would govern socialist agriculture, by contrast with its capitalist form. This complements and continues a central theme of Marx's early writings: a critique of regimes of private property in terms of the estrangement they presuppose and reimpose between humans and the natural world. An external, instrumental relation between humans and the natural conditions, contexts, and subjects of their life activity displaces an orientation to nature in which such activity is a source of intrinsic aesthetic, intellectual, and spiritual fulfillment. Communism will restore to the human species these lost dimensions of their relationship to their nonhuman environment.

An Ecological Crisis Tendency of Capitalism?

I now turn to some brief indications of how Marx's account of specifically capitalist production might be revised in such a way as to provide a powerful explanation of the peculiar liability of this economic form to generate crises of an ecological kind. The argument as developed so far is sufficient for us to see in very general terms why this might be. The dominant labor processes of industrial capitalism, which have transformative intentional structures, are, as we have see, liable to sustain spontaneous ideologies in their economic agents that exclude or occlude their dependence upon contextual conditions and limits. The conceptualizations of the political economists, including Marx, tend simply to reflect the forms of calculation of the economic agents themselves in this respect. Moreover, the value-maximizing intentional structure that is superimposed upon and predominates over the productive intentional structure *intensifies* the latter's insensitivity to material conditions, resources, and limits by its very indifference to the concrete character of the process. Insofar as the expansive dynamic of capital accumulation also requires the production of use values on an ever-expanding scale, it follows that the intrinsic dynamic of capitalist production is a tendency to exceed its extrinsic conditions of sustainability, and, moreover, to do this in ways that are excluded or occluded *by the forms of calculation available to economic agents.*

Even if these forms of calculation were to *become* available, through a shift in the ideology of economic agents, the constraints implicit in their economic situation would preclude any significant shift to an "ecoregulatory" orientation in practice.

This is, of course, a highly schematic and abstract specification of a distinctively capitalist ecological crisis-generating mechanism. To render it more specific and concrete would require us to follow Marx himself in considering the conditions for the reproduction not individual illustrative examples of capitalist production, but of the total social capital. This is Marx's concern in volume 2 of *Capital,* and what he has to say there is potentially very illuminating. As Marx notes, once we consider individual capitalist processes of production in the context of the circulation of the total social capital, conditions previously *assumed* as "given" now must be recognized as problems requiring theorization:

> In both the first and the second Parts it was always only a question of some individual capital, of the movement of some individualized part of capital.
>
> However the circuits of the individual capitals intertwine, presuppose and necessitate one another, and form, precisely in this interlacing, the movement of the total social capital.[21]

Marx's theorization of these relations of mutual presupposition and necessitation of individual capitals involves him in a direct recognition of the pertinence of broad categories of *use values* in the products and materials employed by these individual capitals. Marx's distinction between Departments I and II (production of means of production and of articles of consumption, respectively) does precisely this. The continuity and scale of capitalist production in one "Department" is dependent upon that in the other, and vice versa. The output of Department II, for example, must be equivalent to the total value of the variable capital (labor power) employed in Departments I and II combined (neglecting, for simplicity's sake, capitalists' consumption). In this way, Marx is able to specify the proportionate allocation so capital and labor to the different branches of production, first, at a constant rate, and then on an expanded scale.

However, although Marx's account does embody a recognition that these reproduction requirements include both "the value as well as the substance of the individual component parts" of productive capital, his representation of these conditions is conducted primarily in value terms. Marx focuses upon the analysis of the conditions under which "the individual capitalist can first convert the component parts of his capital into money by the sale of his commodities, and then reconvert them into productive capital by renewed purchase of the elements of production in the commodity-market."[22] clearly, the *immediate* conditions of this possibility include sufficient means of exchange (money) and appropriately proportionate prior allocations of capital and labor across the different branches of production that supply these particular elements of production. Equally clearly, among the *mediate* condition of this possibility are the quantitative proportions of the use-value outputs of those labor processes that appropriate energy, raw materials, and means of subsistence from nature. Marx does not deny this. Indeed, he comments approvingly on Quesnay's statement of it: "The economic process of reproduction, whatever may be its specific social character, always becomes intertwined in this sphere (agriculture) with a natural process of reproduction. The obvious conditions of the latter throw light on those of the former, and keep off a confusion of thought which is called forth by the mirage of circulation.[23]

Notwithstanding Marx's explicit recognition that the whole immense intertwined process of circulation of the total social capital remains bound to its naturally given conditions, and to the labor processes of primary acquisition, he does not pursue the further implications this thought might have. Among these implications, especially if combined with the rectifications I have proposed in Marx's concept of the labor process itself, are a number of insights into the ecological crisis-generating tendencies of capitalist accumulation. First, it becomes possible to perceive in crises of disproportion the mediated and displaced manifestations of crises of an ecological nature whose source is located in those labor processes such as extraction and ecoregulation that are at the "interface" between the total social capital and its natural preconditions. Second, it becomes possible to recognize those branches of "production" (energy generation, the extractive industries, agriculture, and forestry) through which the primary appropriation of nature is conducted as economic loci that focus and concentrate the generalized tendency of capitalist production to exceed its natural limits. They are, so to speak, "pressure points" toward which the ever-growing material requirements of all other social practices are conducted and through which they must flow. Third, it is, as we have seen, precisely in these practices that the intentional structures and forms of calculation of value maximization and transformative action are most severely inappropriate to the sustainability of the practices concerned. Perhaps this is why it was almost exclusively with respect to agriculture that Marx was able to recognize, if only descriptively, the tendency of capitalism to destroy its own natural conditions of possibility:

> Capitalist production . . . disturbs the circulation of matter between man and the soil, i.e. prevents the return to the soil of its elements consumed by man in the form of food and clothing, it therefore violates the conditions necessary to lasting fertility of the soil. . . . Moreover, all progress in capitalistic agriculture is a progress in the art, not only of robbing the labourer, but of robbing the soil, all progress in increasing

the fertility of the soil for a given time, is a progress towards ruining the lasting sources of that fertility.[24]

The hope, finally, is that by beginning to identify and characterize the mechanisms of ecological crisis generation we can ultimately move beyond the methodological weakness of much contemporary ecological analysis that operates by groundlessly extrapolating the mere empirical trends.

SOME CONCLUDING RESERVATIONS

The above analyses are, of course, very partial and abstract. They focus upon labor processes in general, and capitalist production in particular. No attempt has been made to identify and describe ecological crisis generation in, for example, state-socialist societies. Nor have I give extended treatment of the ecological implication of the forms and dynamics of material consumption, important as these are for any political strategy that incorporates an ecological perspective.[25] Nor, indeed, have I considered the powers of economic intervention of capitalist states, in their role as global regulators of the conditions of capitalist production and reproduction. If capitalist economies have intrinsic ecological crisis-generating mechanisms, it remains to be seen whether these crises, along with the more widely recognized forms of economic crisis, can be effectively managed by way of legislation and state interventions. It can also plausibly be argued that some of the great ecological dangers now confronting us are neither direct nor mediated effects of economic relations and dynamics, but arise from relatively autonomous strategic and military policies of nation-states and alliances. Finally, my analysis contributes to addressing, but by no means fully confronts, the most distinctive and dangerous feature of our contemporary ecological crisis: its *global* character. The question is not solely one of identifying the ecological crisis tendencies of specific modes of social and economic life. It is the further, and almost unimaginably complex, one of interpreting the combined and overdetermined interactions of these diverse mechanisms at the level of the ecosphere itself. More, even, than this, "the point is to change it."

NOTES

1. K. Marx, *Capital* (London: Progress, 1961), 1:178.
2. Ibid., 1:180.
3. Ibid., 1:179.
4. Ibid., 1:180.

5. Something already recognized by Adam Smith, who noted that "the most important operations of agriculture seem intended not so much to increase . . . as to direct the fertility of nature towards the production of the plants most profitable to man. . . . Planting and tillage frequently regulate more than they animate the active fertility of nature; and after all their labour, a great part of the work always remains to be done by her." Quoted by Marx in *Capital* (London: Progress, 1970) 2:365.

6. As will become clear later, the claim I am making here is solely that these irrationalities can be explained in terms of the contradictory intentional structure of specifically capitalist production. It is *not* claimed that these kinds are *peculiar* to capitalism.

7. The seminal work for the contemporary debate on the labor process is, of course, Harry Braverman's *Labor and Monopoly Capital* (New York: Monthly Review Press, 1974). The shift of emphasis to which I refer is evident from page 1, where Braverman begins: "All forms of life sustain themselves on their natural environment; thus all conduct activities for the purpose of appropriating natural products to their own use," but quickly goes on to say: "However, what is important about human work is not its similarities with that of other animals, but the crucial differences that mark it *as the polar opposite*" (emphasis added). A valuable collection of essays on issues arising from Braverman's "deskilling" thesis is S. Wood, ed., *The Degradation of Work?* (London: Hutchinson, 1982). See also P. Thompson, *The Nature of Work: An Introduction to Debates on the Labour Process* (London: Basingstoke Macmillan, 2nd ed., 1989) and D. Knights, H. Willmott, and D. Collinson, eds., *Job Redesign* (Aldershot, U.K.: Gower, 1985). In an interesting commentary on an alleged dead end to the labor process debate ("A Labour Process to Nowhere" [*New Left Review, 165*, September–October 1987, 34–35]), S. Cohen has argued for a reintegration of studies of capitalist labor processes with an understanding of their aspect as social–relational processes of valorization and exploitation. While this comment has some purchase on the recent directions of the debate, it does not call into question the still more pervasive occlusion of the aspects of labor processes as forms of appropriation of nature.

8. The literature of the "domestic labor debate" is very extensive. An excellent overview is to be found in M. Barrett, *Women's Oppression Today: Problems in Marxist Feminist Analysis* (London: Verso, 1980), chap. 5, esp. 173–180. See also my own comments in *The Rise and Fall of Structural Marxism* (London: Macmillan, 1984), 135–139.

9. A. Gorz, *Ecology as Politics* (London: Pluto, 1980), 20–28.

10. Here, as elsewhere, I am using the "productive/nonproductive" distinction to discriminate between *intentional structures* of labor processes, not in the more technical sense in which it is commonly used in Marxist economic debate.

11. See G. A. Cohen, *Karl Marx's Theory of History: A Defence* (Oxford: Oxford University Press, 1978), chap. 7.

12. Marx, *Capital* (London: Progress, 1962), 3:254.

13. Engels, "Socialism: Utopian and Scientific," in K. Marx and F. Engels, *Basic Writings on Politics and Philosophy*, ed. C. S. Feuer (London: Fontana/Collins, 1984), 149–150.

14. In the original version of this article, pp. 56–60.

15. This proposal corresponds closely to W. H. Matthews's important contribution "The Concept of Outer Limits," in *Outer Limits and Human Needs*, ed. W. H. Matthews (Uppsala: Almqvist and Wiksell, 1976): "The limits are not, in all cases—not even in most cases—explicit, predictable discrete thresholds which if exceeded produce catastrophic results, regardless of how they are approached. The mental image should not be one of the edge of a cliff where a single additional step plunges one to the depths below. The concept is much more complex and requires full consideration of the role man plays in setting the limits, since these are determined in two ways: (a) by the quantity of existing resources and the laws of nature; but also (b) by the way man conducts his activities with respect to his natural situation" (16–17).

16. This proposal, though conceptualized differently, is very much in line with Timpanaro's comments on the abandonment by contemporary Marxism of the materialist thesis of the dependence of man on nature: "The position of the contemporary Marxist seems at times like that of a person living on the first floor of a house, who turns to the tenant of the second floor and says: 'You think you're independent, that you support yourself by yourself? You're wrong! Your apartment stands only because it is supported on mine, and if mine collapses, yours will too,' and on the other hand to the ground-floor tenant: 'What are you saying? That you support and condition me? What a wretched illusion! The ground floor exists only in so far as it is the ground floor to the first floor'" (see his *On Materialism* [London: New Left Books, 1975], 44).

17. See T. Benton, "Natural Science and Cultural Struggle: Engels on Philosophy and the Natural Sciences," in *Issues in Marxist Philosophy*, Volume 2, eds. J. Mepham and D.-H. Ruben (Brighton, U.K.: Harvester, 1979), 2.

18. From the "Part Played by Labour in the Transition from Ape to Man," in Marx and Engels, *Selected Works* (London: Lawrence and Wishart, 1968), 361–362. It should not be forgotten that one of Engels's earliest works is, in effect, a denunciation of the environmental consequences of capitalist industrialization; see his "The Condition of the Working-Class in England," in Marx and Engels, *Collected Works* (London: Lawrence and Wishart, 1975), 4.

19. Marx, *Capital*, 3:820.

20. Ibid., 1:305–306; emphasis added.

21. Ibid., 2:357. Of course, I have only scratched the surface of what Marx has to say on the conditions of reproduction if the total social capital. My guess, here, is that a fresh look at the debate arising from Rosa Luxemburg's famous use of Marx's reproductive schemes to develop a hypothesis of necessary external limits to capitalist accumulation would yield some interesting insights for an ecological critic. See R. Luxemburg and N. Bukharin, *Imperialism and the Accumulation of Capital*, ed. K. Tarbuck (London: Allen Lane/Penguin, 1972).

22. Marx, *Capital*, 2:397.

23. Ibid., 2:363.

24. Ibid., 1:505–506.

25. A critical discussion of CPE's assumption of "limitless desires" as subjective conditions for capital accumulation (exemplified in the passage from

Ricardo quoted in the original version of this article) would be a useful starting point. It would have to tackle the question of how to conceptualize human needs, and the limits to the commodification of the means of satisfying them. Important recent work that takes up the direction of such a theory of needs includes: W. Leiss, *The Limits of Satisfaction* (London: Marion Boyars, 1978); K. Soper, *On Human Needs* (Brighton, U.K.: Harvester, 1981); and L. Doyal and I. Gough, "A Theory of Human Needs," *Critical Social Policy*, No. 10, Summer, 1984. See also my own suggestions for a naturalistic but nonreductionist theory of needs, in "Humanism vs. Speciesism," *Radical Philosophy, 50,* Autumn, 1988, esp. 13–15; and my subsequent development of this argument in *Natural Relations: Ecology, Animal Rights, and Social Justice* (London: Verso, 1993).

PART III

The Second Contradiction of Capitalism

INTRODUCTION
TO PART III

Part III of this collection is devoted to James O'Connor's seminal "The Second Contradiction of Capitalism" and a selection of the published responses to it. His paper formed the theoretical introduction to the first issue of *Capitalism, Nature, Socialism,* and has inspired a considerable subsequent debate, much of it in the pages of *CNS.* A future volume in this series will be devoted to O'Connor's work, including responses to his critics, and further revisions and development of his ideas. In this volume, we have confined ourselves to reprinting the original paper, retitled and with minor textual corrections, together with a small selection of critical pieces from *CNS.* The main principle of selection here has been to include those contributions to the debate that challenge the core concept of the second contradiction. Other contributions will be subjects of discussion in the forthcoming volume.

O'Connor offers what is perhaps the most thoroughly worked out and systematic development of an ecological Marxism so far. In effect, he provides us with a general theory of capitalist development and its potential for socialist transcendence that fully incorporates ecological crises and social movements. The argument is pitched at a high level of abstraction, while explicitly recognizing the significance of local contexts and contingencies in the actual working out of the abstract tendencies and contradictions outlined in the theory, and calling for empirical work to test and develop these ideas.

The central line of argument is that the classical Marxist account of the central contradiction of capitalism—between forces and relations of production—needs to be supplemented with a recognition of a second contradiction, between forces (and relations) of production and conditions of production. The first contradiction generates crises of overproduction, through which forces and relations of production are restructured in the direction of a growing socialization of the forces of production. The first contradiction engenders a labor movement, which constitutes a social barrier to capital accumulation and which may provide the agency for a transition to socialism. This latter is not guaranteed in advance, but is at least rendered imaginable by the dereifying effects of capitalist crisis, and by the increasingly socialized character of capitalist production itself.

The second contradiction is distinct from, but analogous to, the first in its tendency to generate crises, to engender a social movement that acts as a barrier to capital accumulation, and to require restructuring of the (provision of) conditions of production in the direction of greater socialization. There may, then, be two distinct routes to socialism: by way of the labor movement, rooted in the first contradiction, and by way of the environmental (and other social) movements, rooted in the second contradiction of capitalism. O'Connor points out that this leaves open for further research the *interaction between* the two contradictions and their further consequences, and the associated relations between labor, environmental, and other "new" social movements. Clearly, some form of anticapitalist alliance or coordination between these distinct movements is envisaged, but relatively little is said about the concrete forms this might take.

Earlier (see Introduction to Part II) I identified four areas of ecological Marxist theoretical work. It seems to me that O'Connor has been relatively little concerned with two of them: the development of an ecological socialist perspective in environmental philosophy and the redefinition of the content of the socialist project (although he makes interesting remarks bearing on these matters in passing). Thus this introductory discussion will focus on the two areas where his contribution has been the most substantial: the reconstruction of the materialist conception of history as an explanatory theory that addresses the ecological dimensions of social life, and the development of new precepts on the question of agency and strategy for change.

Some contributors to the debate on the second contradiction expressed reservations about the alleged narrowness of its focus: in particular, the location of the primary cause of all forms of environmental degradation in a single contradiction of capitalist economic relations. In Toledo's view, for example, these problems should be understood more broadly as symptomatic features of a whole civilization or mode of life. Marx and the Marxian tradition should be explored as a basis for a whole new political philosophy for the green movement. Toledo, like other writers in this collection, points to the environmentally destructive character of the ex-socialist regimes. He also notes environmental successes in market economies. So is capitalism the source of *all* environmental evil?

In defense of O'Connor's position, it could be argued that a demonstration that ecological crises are endemic features of capitalist societies does not commit him to the view that all ecological problems are rooted in capitalism. A fully *comprehensive* account of the world's environmental problems would clearly require much more than a theory of capitalist ecological crisis, but it could still be argued, given the global preeminence of capitalist economic relations, that such a theory must form an *essential contribution* to this task. However, an important consequence of the point made by Toledo is that the transcendence of capitalist relations would not of itself guarantee ecological harmony. To think coherently about how a postcapitalist society might resolve the ecological problems bequeathed by capitalist and state-socialist regimes *would* require a more broadly based theory of human-historical ecology.

Insofar as Toledo also appears to be criticizing O'Connor's thesis for its economic determinism, another line of defense seems possible. O'Connor is clearly working within a conception of capitalism that does not reduce it to a set of economic relations. For O'Connor, cultural processes are at work in the forces and relations of production, and the definition of "conditions of production" that he gives clearly includes a wide range of social and political relations, including family forms, urban processes, means of communication, and the institutions of the state that are involved in restructuring the conditions of production.

However, it still seems to me that there is some validity in this criticism of the second contradiction thesis. It can, for example, reasonably be argued that state-policies may be an autonomous source of major environmental problems that have at most very indirect relationships to capital accumulation. One example would be the economically irrational development of the nuclear power and reprocessing industries linked to military requirements for plutonium. Also, within what would generally be regarded as economic relations, O'Connor's focus on the forces and conditions of production as sources of ecological crisis leaves out of account environmental consequences of the *consumption* of commodities: the most obvious example here is private car use and associated infrastructural and demographic causes of environmental destruction, atmospheric pollution, and so on. This is not to say, of course, that consumption should be considered in isolation from production. However, it is crucially important, especially for understanding the politics of the environmental movements, to consider consumption as a distinct "moment" of socioeconomic activity.

However, there is a third, and perhaps more serious limitation, in an *exclusive* focus on the relationship between capitalist production and its various conditions. This is not so much because ecological destruction and degradation have significant other causes, but more because the ecological consequences of the second contradiction have a significance that is much wider than their effect on capital accumulation. The whole range of popular pleasures and practices that go to make up "everyday life" outside the sphere of production also, like production itself, have their "conditions": streets that are safe enough, quiet enough, and free enough of pollution for kids to play outside, for women to walk freely, and for neighbors to meet and talk; access to the local countryside; public spaces for meetings, celebrations, performances, and festivals; and so on. Sometimes these conditions are *identical* with conditions of production, but they may also be *connected* to conditions of production by way of more or less complicated pathways of ecological degradation. So the tendency of capital to undermine its own conditions *may also* entail a tendency to undermine the conditions of *other* nonproductive social practices. It is often these more indirect and attenuated effects of the second contradiction on peoples' quality of life that form the "raw materials" of environmental issues and struggles.

This is not, of course, denied in anything that O'Connor says (indeed, it is close to what he does say in his concluding section), but it is worth spelling out the argument, since it suggests that even where *causes* lie in capitalist production, social movements that arise in relation to *effects* may assign meanings and develop purposes that do not relate to the perspective of "the unity of social labor." Also, since different social classes and other social categories are affected in different ways, and have different material and cultural resources with which to respond to processes of environmental degradation, the result is likely to be a highly fluid, fragmented, and often politically divided set of responses, rather than a single, anticapitalist movement. This is one reason why Michael A. Lebowitz's proposal to represent both the "first" and the "second" contradictions of capitalism as but forms of a single contradiction (between the "needs" of capital and the "needs" of humans) is likely to be a step in the wrong direction. Either Lebowitz's concept of the "modern multifarious proletariat" does acknowledge this sociopolitical diversity and ambivalence of the social movements, in which case the concept "proletariat" is vastly overextended, or it does not, in which case it would lead to strategic errors of a potentially disastrous nature.

However, if we set aside these arguments about whether the right starting point for an ecologically revised Marxism is, after all, the postulated second contradiction, the next set of questions has to do with how this second contradiction is conceptualized, and what it is supposed to explain. Since this second contradiction is represented as a relation between capitalist forces (and relations) and the conditions of production, it will first be necessary to consider this distinction itself. It will be recalled, from my own contribution to Part II of this book, that Marx himself, though he did distinguish "conditions" of production, was often inclined to include them within the broader category of "instruments" of production, that is, to include them among the "forces." There is a similar ambiguity about human labor, labor power, and the person of the laborer. Do these belong to the "forces" or to the "conditions" of production? O'Connor provides a valuable classification, and also a theoretically powerful definition, of the "conditions." Following Marx, he distinguishes among the conditions of production, "external physical conditions" (including conditions that would commonly be regarded as features of the "natural environment": ecosystems, soil, air, water, etc.), "labor power" (including workers as biological organisms, mental and physical health, etc.), and the "communal, general conditions of social production" (including means of communication, transport, infrastructure, etc.).

The definition that brings these diverse categories together as "conditions of production" is that capital treats them *as if* they were commodities, even though they are not produced and reproduced capitalistically. This latter fact about them implies that the market cannot be relied upon to regulate their provision, or repair, replacement, modification, and the like in response to demand. Some noneconomic intervention (O'Connor assumes this is either the state or "capitals acting as if they are the state") is therefore required to secure provision of appropriate conditions for capital accumulation. It is for this reason that crises produced by the contradiction between forces and conditions of production "dereify" and so politicize the restructuring of the conditions of production. The state, or, more broadly, sociopolitical processes "mediate" between capital and nature.

This is the core of O'Connor's argument, and it is a very powerful one. However, there are some unresolved problems in the way it is formulated. One problem is that it is not generally true that the conditions of production are treated by capital "as if they were commodities." This is, indeed, true of labor power and O'Connor's framework could well provide insights into family policy, health care, educational and training provision, industrial injuries compensation, and so on in terms of the second contradiction. However, many natural environmental and infrastructure conditions of production are not treated by capital as commodities, but rather as "free goods." The main strategy of neoclassical environmental economics, and of a large part of the reform wing of the environmental movement, is precisely to *require* capital to treat these conditions as commodities, to internalize them as part of their cost structure. Clearly, as O'Connor suggests, this growing capitalization of nature requires legal and political regulation, but it is not obvious that this is qualitatively different from any other sphere of capitalist activity in that respect. Moreover, there is a growing tendency for reformist environmentalism to provide an ideological *legitimation* for such extensions of state power as are needed to increase the commodification of nature and infrastructural conditions (e.g., privatization of waste disposal, highway tolls, privatization and commodification of water supplies, patenting of genetic materials, taxing pollution and use of nonrenewable resources, etc.).

The paradox here is that what O'Connor's analysis reveals as the central cause of environmental degradation under capitalism is viewed by "ecologically enlightened" capital and by environmental reformism as the key to its solution. The air of paradox dissolves, of course, when we recognize that O'Connor is not so much explaining environmental *degradation* as environmental *crisis*. The commodification of nature is capital's (political) response to ecological degradation that has become an obstacle to further capital accumulation. Still, the position remains problematic for two reasons. The first is that O'Connor seems to theorize the necessity for capital to restructure the conditions of production by way of crises as a consequence of capitalism's general tendency to destroy its own conditions. However, there is a further, and perhaps more fundamental, dynamic at work. This is the development of the forces of production themself. Scientific advances harnessed to capitalist technological innovations opens up new domains of nature to economic exploitation, and successive waves of capitalist expansion carried by these new domains of exploitation themselves require new forms of capitalization of nature, quite independently of their ecologically destructive side effects. The bids by pharmaceutical transnationals to buy up, for commercial exploitation, vast areas of rain forest and conflicts over the legal right to patent genetically engineered organisms are significant examples associated with the production-condition requirements of the capitalist exploitation of the new biotechnologies. Thus any restructuring of the provision of conditions of production has to be understood in terms of more or less complicated combinations of *both* shifts in the scientific and technical components of the forces of production *and* responses to the environmental consequences of past production.

The second reason why the position remains problematic is that it is now unclear why the term "contradiction" is appropriate. O'Connor has shown (and my argument here reinforces this) that capitalist development requires recurrent restructuring of the ways in which conditions of production are provided. He has also shown that these processes require some form of extraeconomic intervention and regulation. However, the claim that there is a "contradiction" between forces and conditions presumably means more than this: that there is a structural fault, intrinsic to capitalism, that is responsible for at least a *tendency* toward transition from this socioeconomic form to some other (i.e., socialism). Now, O'Connor's characterization of the first contradiction is already highly qualified in this respect. The first contradiction engenders the labor movement which has the *potential* to be the agent of a transition to socialism, but there is no necessity about this. The effects of the first contradiction are to dereify capitalist relations, and in socializing capitalist production to make social-ism "imaginable."

One could argue about whether a "potential" that is, in this weakened form, little more than a formal *possibility* is sufficient grounding for the concept "contradiction." However, this would be a diversion from the central issue which is, it seems to me: Is there enough of an analogy between the second contradiction and the first contradiction, even given O'Con-nor's dilution of traditional Marxism's version of it, to justify giving it this status? Other contributors to the second contradiction debate, notably Lebowitz, insist that there *are* significant disanalogies, which O'Connor overlooks in his development of a schematic parallel between the two "contradictions." Here, there is space only to mention a few differences. First, as we have already noticed, restructuring of the provision of a whole range of what were formerly "free goods" so that they can be treated by capital "as if they were commodities" has often taken the form either of privatization of public utilities or of an extension of private property rights. Though these processes are themselves often politically contentious, they frequently lead to new forms of institutionalization (e.g., private-sector monopolies) which *reduce* their political accessibility, and also constitute a shift *away* from "socialization."

Second, as we have also seen, environmental movements do not necessarily or universally resist theses changes. Sometimes they are actively demanded by reform environmentalists since they generally entail an internalization of environmental externalities that would otherwise remain uncontrolled. In this respect, parts of the environmental movement, at least, may be acting against individual capitals, but at the same time furthering the longer term interests of capital-in-general by legitimating new forms of provision of conditions for further capital accumulation. So, while it may be clear how the labor movement forms a "social barrier" to capital, it is not clear that this is true of the environmental movement *in all of its manifestations*. This, of course, takes us back to my earlier point that we should expect an irreducible plurality and political ambivalence among the social groupings and organizations that form up under the banner of "environmentalism." So, the thesis that O'Connor proposes, albeit tenta-tively, that there may be two, parallel paths to socialism, one via the labor movement, the other by way of the environmental and other social movements, seems to be unfounded. The case for calling the second contradiction a "contradiction" at all remains to be made. This seems to be one of Andriana Vlachou's conclusions, when she says "I think there is no a priori tendency for capitalism to produce environmental crises."

Third (and relatedly), the labor movement could be said to be engendered by the contradiction between forces and relations of production in ways that do not apply to the "engendering" of the environmental and other "new" social movements by the alleged contradiction between the forces and the conditions of production. This is clearly acknowledged by O'Connor himself when he says "most struggles have strong, particularistic 'romantic anticapitalist' dimensions." However, as the contributions by Vlachou and Parlato and Ricoveri suggest, the differences go deeper than this. Forms of (class) identity, community, and organization have historically emerged on the basis of shared recognitions of a common relationship to the relations of production: capitalist property. Of course, these "shared recognitions" have been less often and less widely "shared" than Marx expected, and less so than many Marxists have assumed. A high degree of particularity and site-specificity has attached to the formation of working-class social movements and political organizations.

However, with respect to the "new" social movements, what we might call their "material grounding" is still more diverse and variable. Several contributions to the second contradiction debate have suggested the focus on "conditions of production" is too narrow for an understanding of the sources of social movements. However, even if we follow O'Connor in understanding them in some sense as "really" engaged in struggles over the protection and/or restructuring of the conditions of capitalist production, even though they may rarely think of *themselves* as doing this, there are still problems for his position. By his own account the concept "conditions of production" is a very wide and heterogeneous category, including both socially provided and naturally presupposed processes, mechanisms, materials, and other conditions. In addition, the ways in which capital destroys and/or seeks to restructure these conditions are very complex, diverse, contingent, and often highly mediated. They impact on the lives of differently socially situated individuals—men and women; ethically segregated populations; occupational categories; infants, children, youths, and the elderly; urban, suburban, and rural dwellers—in profoundly different ways. All of these will be shaped by class divisions but cannot be *reduced* to a class-specific measure of environmental vulnerability. Perhaps this is what O'Connor means when he says these issues are "class issues, even though they are *more* than class issues." However, these considerations seem to me to suggest that there may just not be the kind of material underpinning of environmentalist calls to unity based on a common relation to capitalist economic relations that gave weight to the idea of labor movement unity.

So an irreducible diversity of values and perspectives among the various social movements may be a permanent strategic reality, not because of the "free-floating" character of the discourses through which social and political identities are secured (as some poststructuralists might argue), but as a result of *substantial differences of life situation* among those affected by what O'Connor calls the "second contradiction." Perhaps we are now in a position to recognize in a way that Marxists—especially those influenced by the Leninist tradition—in the past have failed to recognize that no single social group can validly claim to represent the interests of all the oppressed, exploited, and excluded. In the words of Parlato and Ricoveri, "No one group, class, or faction can possibly gather the strength and will to confront global capital on the nature, labor, gender, place, and antiracist fronts simultaneously . . . yet we need to remember that 'everything is fresh off the machine' and that we are still at the beginning of a new phase."

Chapter 9

THE SECOND
CONTRADICTION
OF CAPITALISM

JAMES O'CONNOR

*Those who insist that [environmental destruction] has
nothing to do with Marxism merely ensure that what
they choose to call Marxism will have nothing to do
with what happens in the world.*
                              —AIDEN FOSTER-CARTER

INTRODUCTION

This chapter expounds the traditional Marxist theory of the contradiction
between forces and relations of production, overproduction of capital and
economic crisis, and the process of crisis-induced restructuring of productive forces and production relations into more transparently social, hence
potentially socialist, forms. This exposition provides a point of departure
for an "ecological Marxist" theory of the contradiction between capitalist
production relations and forces and the *conditions of production*, underproduction of capital and economic crisis, and the process of crisis-induced
restructuring of production conditions and the social relations thereof,
also into more transparently social, hence potentially socialist, forms. In

197

short, there may be not one but two paths to socialism in late capitalist society.

While the two processes of capital overproduction and underproduction are by no means mutually exclusive, they may offset or compensate for one another in ways that create the *appearance* of relatively stable processes of capitalist development. Study of the combination of the two processes in the contemporary world may throw light on the decline of traditional labor and socialist movements and the rise of "new social movements" as agencies of social transformation. In ways similar to how traditional Marxism illuminates the practices of traditional labor movements, it may be that "ecological Marxism" throws light on the practices of new social movements. Although ecology and nature; the politics of the body, feminism, and the family; and urban movements and related topics are usually discussed in post-Marxist terms, the rhetoric deployed in this chapter is self-consciously Marxist and designed to appeal to Marxist theorists and fellow travelers whose work remains within a "scientific" discourse, and hence those who are least likely to be convinced by post-Marxist discussions of the problem of capital's use and abuse of nature (including human nature) in the modern world. However, the emphasis in this chapter on a political, economic, "scientific" discourse is tactical, not strategic. In reality, more or less autonomous social relationships, often noncapitalist or anticapitalist, constitute "civil society," which needs to be addressed on its own practical and theoretical terms. In other words, social and collective action is not meant to be construed merely as derivative of systemic forces, as the last section of this chapter hopefully will make clear.

In 1944, Karl Polanyi published his masterpiece, *The Great Transformation,* which discussed the ways in which the growth of the capitalist market impaired or destroyed its own social and environmental conditions.[1] The subject of the ecological limits to economic growth and the interrelationships between development and environment was reintroduced into Western thought in the 1960s and early 1970s. The results have been mixed and highly dubious. Polanyi's work remains a shining light in a heaven filled with dying stars and black holes of bourgeois naturalism, neo-Malthusianism, Club of Rome technocratism, romantic deep ecologyism, and United Nations one-worldism.[2] Class exploitation, capitalist crisis, uneven and combined capitalist development, national independence struggles, and so on are missing from these kinds of accounts. The results of these and many—if not most—other efforts to discuss the problem of capitalism, nature, and socialism wither on the vine because they fail to focus on the nature of specifically capitalist scarcity, that is, on the process whereby capital is its own barrier or limit because of its self-destructive forms of proletarianization of human nature, and appropriation of labor and capitalization of external nature.[3] The usual approaches to the problem—the identification of "limits to growth" in terms of "resource scarcity," "ecological fragility," "harmful industrial technology," "destructive cultural values," "tragedy of the commons," "overpopulation," "wasteful consumption," "production treadmill," and so on either ignore or mangle Marx's theories of historically produced forms of nature and capitalist accumulation and development.

This is not surprising since Marx wrote little pertaining to the ways that capital limits itself by impairing its own social and environmental conditions, hence increasing the costs and expenses of capital, thereby threatening capitals" ability to produce profits, that is, threatening economic crisis. He also wrote little or nothing about the effects of social struggles organized around the provision of the conditions of production on the costs and expenses and variability of capital. Nor did he theorize the relationship between social and material dimensions of production conditions, excepting his extended discussion of ground rent (i.e., social relations between landed and industrial capital and material and economic relations between raw materials and industrial production). Marx was, however, convinced of at least three things. The first was that deficiencies of production conditions or "natural conditions" ("bad harvests") may take the form of economic crisis.[4] Second, he was convinced of the more general proposition that some barriers to production are truly external to the mode of production ("the productiveness of labour is fettered by physical conditions"),[5] but that in capitalism these barriers assume the form of economic crisis.[6] Put another way, some barriers are "general" not "specific" to capitalism. What is specific is the way these barriers assume the form of crisis. Third, Marx believed that capitalist agriculture and silviculture are harmful to nature, not only that capitalist exploitation is harmful to human labor power.

In sum, Marx believed that capitalist farming (for example) ruined soil quality. He was also clear that bad harvests take the form of economic crisis. However (although he did state that a rational agriculture is incompatible with capitalism),[7] he never considered the possibility that ecologically destructive methods of agriculture might raise the costs of the elements of capital, which, in turn, might threaten economic crisis of a particular type, namely, underproduction of capital.[8] Put another way, Marx never put two and two together to argue that "natural barriers" may be capitalistically produced barriers, that is, a "second" capitalized nature.[9] In other words, Marx hinted at but did not develop the idea that there may exist a contradiction of capitalism that leads to an "ecological" theory of crisis and social transformation.

TWO KINDS OF CRISIS THEORY

The point of departure of the traditional Marxist theory of economic crisis and the transition to socialism is the contradiction between capitalist productive forces and production relations.[10] The specific form of this contradiction is between the production and realization of value and surplus value, or between the production and circulation of capital. The agency of socialist revolution is the working class. Capitalist production relations constitute the immediate object of social transformation. The site of transformation is politics and the state and the process of production and exchange.

By contrast, the point of departure of an "ecological Marxist"[11] theory of economic crisis and transition to socialism is the contradiction between capitalist production relations (*and* productive forces) and the *conditions* of capitalist production, or "capitalist relations and forces of social reproduction."[12]

Marx defined three kinds of production conditions. The first is "external physical conditions,"[13] or the natural elements entering into constant and variable capital. Second, the "labor power" of workers was defined as the "personal conditions of production." Third, Marx referred to "*the communal, general conditions of social production,* for example, "means of communication."[14]

Today "external physical conditions" are discussed in terms of the viability of ecosystems; the adequacy of atmospheric ozone levels; the stability of coastlines and watersheds; soil, air, and water quality; and so on. "Labor power" is discussed in terms of the physical and mental well-being of workers, the kind and degree of socialization, toxicity of work relations and the workers" ability to cope, and human beings as social productive forces and biological organisms generally. "Communal conditions" are discussed in terms of "social capital," "infrastructure," and so on. Implied in the concepts of "external physical conditions," "labor power," and "communal conditions" are the concepts of space and "social environment." We include as a production condition, therefore, "urban space" ("urban capitalized nature") and other forms of space that structure and are structured by the relationship between people and "environment,"[15] which, in turn, helps to produce social environments. In short, production conditions include commodified or capitalized materiality and sociality excluding commodity production, distribution, and exchange (strictly defined) themselves.

The specific form of the contradiction between capitalist production relations (and forces) and production conditions is also between the production and realization of value and surplus value. The agency of social transformation is "new social movements," social struggles including struggles within production over workplace health and safety, toxic waste production and disposal, the use of urban space, and the like. The social relationships of reproduction of the conditions of production (e.g., state and family as structures of social relations and also the relations of production themselves insofar as "new struggles" occur within capitalist production) constitute the immediate object of social transformation. The immediate site of transformation is the material process of reproduction of production conditions (e.g., division of labor within the family, land use patterns, education, etc.) and the production process itself, again insofar as new struggles occur within the capitalist workplace.

In traditional Marxist theory, the contradiction between production and realization of value and economic crisis takes the form of a "realization crisis," or overproduction of capital. In ecological Marxist theory, economic crisis assumes the form of a "liquidity crisis," or underproduction of capital. In traditional theory, economic crisis is the cauldron in which capital restructures productive forces and production relations in ways that make both more transparently social in form and content, for example, indicative planning, nationalization, profit sharing, and the like. In ecological Marxism, economic crisis is the cauldron in which capital restructures the conditions of production also in ways that make them more transparently social in form and content, for example, permanent-yield forests, land reclamation, regional land use and/or resource planning, population policy, health policy, labor market regulation, toxic waste disposal planning, and so on.

In traditional theory, the development of more social forms of productive forces and production relations is regarded as a necessary but not sufficient condition for the transition to (productivist) socialism. In ecological Marxism, the development of more social forms of the provision of the conditions of production is also a necessary but not sufficient condition for (ecological) socialism. "Ecological socialism" would be different than that imagined by traditional Marxism, first, because from the perspective of the "conditions of production" most struggles have strong, particularistic "romantic anticapitalist" dimensions, that is, they are "defensive" rather than "offensive," and, second, because it has become obvious that much capitalist technology, forms of work, and the like, including the ideology of material progress, have become part of the problem not the solution. In sum, there may be not one but two paths to socialism, or, to be more accurate, two tendencies that together lead to increased (albeit historically reversible) socialization of productive forces, production relations, conditions of production, and social relations of reproduction of these conditions.

THE TRADITIONAL MARXIST ACCOUNT
OF CAPITALISM AS A CRISIS-RIDDEN SYSTEM

In traditional Marxism, the contradiction between the production and circulation of capital is "internal" to capitalism because capitalist production is not only commodity production but also production of surplus value (i.e., exploitation of labor). It is a valorization process in which capitalists extract not only socially necessary labor (labor required to reproduce constant and variable capital) but also surplus labor from the working class. Everything else being the same,[16] any given amount of surplus value produced and/or any given rate of exploitation will have the effect of creating a particular shortfall of commodity demand at market prices. Or, put the opposite way, any particular shortage of commodity demand presupposes a given amount of surplus value produced and/or a given rate of exploitation. Further, the greater the amount of surplus value produced and/or the higher the rate of exploitation, the greater the difficulty of realizing value and surplus value in the market. The basic problem of capitalism is: Where does the extra commodity demand that is required to buy the product of surplus labor originate? Time-honored answers include capitalist class consumption; capital investment that is made independently of changes in wage advances and consumer demand; markets created by these new investments; new investment, consumption, or government spending financed by expanded business, consumer, or government credit; the theft of markets of other capitals and/or capitals in other countries; and so on. However, these "solutions" to the problem of value realization (that of maintaining a level of aggregate demand for commodities that is sufficient to maintain a given rate of profit without threatening economic crisis and the devaluation of fixed capital) turn into other kinds of potential "problems" of capitalism. Capitalist consumption constitutes an unproductive use of surplus value, as does the utilization of capital in the sphere of circulation with the aim of selling commodities faster. New capital investment may expand faster than, or independently of, new consumer demand with the result of increasing chances of a disproportionality crisis or a more severe realization crisis in the future. While a well-developed credit system can provide the wherewithal to expand commodity demand independent of increases in wages and salaries, the expansion of consumer demand based on increases in consumer or mortgage credit greater than increases in wages and salaries threatens to transform a potential crisis of capital overproduction into a crisis of capital underproduction. Moreover, any expansion of credit creates debt (as well as assets), financial speculation, and instabilities in financial structures, and thus may threaten a crisis in the financial system. The theft of markets from other capitals implies the concentration and/or centralization of capital, and hence a worsening of the problem of realization of value in the future and/or social unrest arising from the destruction of weaker capitals, or political instability, bitter international rivalries, protectionism, and war. In sum, economic crisis can assume varied forms besides the traditional "realization crisis," including liquidity crisis, financial crisis or collapse, fiscal crisis of the state, and social and political crisis tendencies. However, whatever the specific forms of historical crises (the list above is meant to be suggestive not exhaustive), and whatever the specific course of their development and resolution, most—if not all—Marxists accept the premise based on the real conditions of capitalist exploitation that capitalism is a crisis-ridden system.

THE TRADITIONAL MARXIST ACCOUNT OF CAPITALISM AS A CRISIS-DEPENDENT SYSTEM AND THE TRANSITION TO SOCIALISM

In traditional Marxism, capitalism is not only crisis-ridden but also crisis-dependent. Capital accumulates through crisis, which functions as an economic disciplinary mechanism. Crisis is the occasion that capital seizes to restructure and rationalize itself in order to restore its capacity to exploit labor and accumulate. There are two general, interdependent ways in which capital changes itself to weather the crisis and resolve it in capital's own favor. One is changes in the productive forces, the second is changes in the production relations. Changes in either typically presuppose or require new forms of direct and indirect cooperation within and between individual capitals and/or within the state and/or between capital and the state. More cooperation or planning has the effect of making production more transparently social, meanwhile subverting commodity and capital fetishism, or the apparent "naturalness" of capitalist economy. The telos of crisis is thus to create the possibility of imagining a transition to socialism.

Crisis-induced changes in productive forces by capitals seeking to defend or restore profits (and exemplified by technological changes that lower unit costs, increase flexibility in production, etc.) have the systematic effect of lowering the costs of reproducing the work force, making raw materials available more cheaply or their utilization more efficient, reducing the period of production and/or circulation, and so on. Whatever the immediate sources of the crisis, restructuring productive forces with the aim of raising profits is a foregone conclusion. Moreover, crisis-induced changes in productive forces imply or presuppose more social forms of production relationships, for example, more direct forms of cooperation within production.[17] Examples of changes in productive forces today, and associated changes in production relationships, include computerized, flexible manufacturing systems and robotics, which are associated with the development of "creative team play" and other forms of cooperation in the workplace and profit sharing, among other changes. And, of course, the greatest productive force is human cooperation, and science or the social production of practical knowledge has become an almost completely cooperative enterprise[18] partly as a result of cumulative historical economic, social, and political crises.

The second way that capital restructures itself is through crisis-induced changes in production relations within and between capital, within the state, and/or between the state and capital, which are introduced with the aim of exercising more control over production, investment, markets, and so on, that is, more planning. Historically, planning has taken many forms (e.g., nationalization, fiscal policy, indicative planning), including, at the political level, fascism, New Dealism, and social democracy. Whatever the immediate sources of crisis, the restructuring of production relations with the aim of developing more control of labor, raw material supplies, and the like, is a foregone conclusion. Furthermore, crisis-induced changes in production relations imply or presuppose more social forms of productive forces, for example, more direct forms of cooperation. Examples of changes in production relations today include "strategic agreements" between high-tech capitals, massive state intervention in financial markets, and centralization of capital via takeovers and mergers. These changes imply sharing or socialization of high-tech secrets and technical personnel, new forms of financial controls, and the restructuring of management and production systems, respectively.

To sum up: Crisis forcibly causes capital to lower costs and increase flexibility and to exercise more control or planning over production and circulation. Crisis causes new forms of flexible planning and planned flexibility (including at the level of state-organized production), which increases the tensions between a more flexible capitalism (usually market-created) and a more planned capitalism (usually state-created). Crisis forcibly makes capital confront its own basic contradiction which may be subsequently displaced to the state, corporate management, and other spheres, when more social forms of productive forces and production relations are introduced, which imply or presuppose one another meanwhile developing independently of one another. In this way, capital itself creates some of the technical and social preconditions for the transition to socialism. However, whether we start from the productive force or from the production relation side, it is clear that technology and power embody one another, and hence that new forms of cooperation hold out only tenuous and ambiguous promises for the possibilities of socialism. For example, state capitalism and political capitalism contain within them socialist forms, but highly distorted ones, which in the course of the class struggle may be politically appropriated to develop less distorted social forms of material and social life. But this is a highly charged political and ideological question. Only in a limited sense can it be said that socialism is imminent in crisis-induced changes in productive forces and production relations. Whether or not these new social forms are imminently socialist forms depends on the ideological and political terrain, degree of popular mobilization and organization, national traditions, and the like, including and especially the particular world economic and political conjuncture. The same cautionary warning applies to the specific forms of cooperation in the workplace that emerge from the crisis, which may or may not preclude other forms that would lend themselves better to socialist practice, which cannot be regarded as some fixed trajectory but is itself an object of struggle, and defined only through struggle.

Nothing can be said a priori about "socialist imminence" except at the highest levels of abstraction. The key point is that capitalism *may* self-destruct or subvert itself when it switches to more social forms of production relations and forces. The premise of this argument is that *any given set of capitalist technologies and work relations is consistent with more than one set of production relations, and that any given set of production relations is consistent with more than one set of technologies and work relations.* The "fit" between relations and forces is thus assumed to be quite loose and flexible. In the crisis, there is a kind of two-sided struggle to fit new productive forces into new production relations and vice versa in more social forms without, however, any "natural" tendency for capitalism to transform itself to socialism. Nationalization of industry, for example, may or may not be a step toward the socialization of industry. It is certainly a step toward more social forms of production and a more specifically political form of appropriation and utilization of surplus value. On the other side, quality circles, work teams, technology sharing, and the like may or may not be a step toward socialism. They are certainly steps toward more social forms of productive forces.

TOWARD AN ECOLOGICAL MARXIST ACCOUNT OF CAPITALISM AS A CRISIS-RIDDEN SYSTEM

The point of departure of "ecological Marxism" is the contradiction between capitalist production relations and productive forces and conditions of production. Neither human labor power, nor external nature, nor infrastructures, including their space/time dimensions, are produced capitalistically, although capital treats these conditions of production *as if* they are commodities or commodity capital. Precisely because they are not produced and reproduced capitalistically, yet are bought and sold and utilized as if they were commodities, the conditions of supply (quantity and quality, place and time) must be regulated by the state or capitals acting as if they are the state. Although the capitalization of nature implies the increased penetration of capital into the conditions of production (e.g., trees produced on plantations, genetically altered species, private postal services, voucher education, etc.), the state places itself between capital and nature, or mediates capital and nature, with the immediate result that the conditions of capitalist production are politicized. This means that whether or not raw materials, labor force, and useful spatial and infrastructural configurations are available to capital in requisite quantities and qualities and at the right time and place depends on the political power of capital, the power of social movements that challenge particular capitalist forms of production conditions (e.g., struggles over land as means of production versus a means of consumption), state structures that mediate or screen struggles over the definition and use of production conditions (e.g., zoning boards), and so on.[19] Excepting branches of the state regulating money and certain aspects of foreign relations (those that do not have any obvious relation to accessing foreign sources of raw materials, labor power, etc.), every state agency and political party agenda may be regarded as a kind of interface between capital and nature (including human beings and space). In sum, whether or not capital faces "external barriers" to accumulation, including external barriers in the form of new social struggles over the definition and use of production conditions (i.e., "social barriers" that mediate between internal or specific and external or general barriers)[20]; whether or not these "external barriers" take the form of economic crisis; and whether or not economic crisis is resolved in favor of or against capital are political and ideological questions first and foremost, and economic questions only secondarily. This is so because production conditions are by definition politicized (unlike production itself); access to nature is mediated by struggles while external nature has no political identity and subjectivity of its own.[21] Labor power alone struggles around the conditions of its own well-being and social environment broadly defined.

An ecological Marxist account of capitalism as a crisis-ridden system focuses on the way that the combined power of capitalist production relations and productive forces self-destruct by impairing or destroying rather than reproducing their own conditions ("conditions" defined in terms of both their social and material dimensions). Such an account stresses the process of exploitation of labor and self-expanding capital, state regulation of the provision of production conditions, and social struggles organized around capital's use and abuse of these conditions. The main question—Does capital create its own barriers or limits by destroying its own production conditions?—needs to be asked in terms of specific use values, as well as exchange value. This is so because conditions of production are not produced as commodities, and thus problems pertaining to them are "site-specific," including the individual body as an unique "site." The question—Why does capital impair its own conditions?—needs to be asked in terms of the theory of self-expanding capital, its universalizing tendencies which tend to negate principles of site-specificity, its lack of ownership of labor power, external nature, and space, and hence (without state or monopolistic capitalist planning) capital's inability to prevent itself from impairing its own conditions. The question—Why do social struggles against the destruction of production conditions (which resist the capitalization of nature, e.g., environmental, public health, occupational health and safety, urban, and other movements) potentially impair capital flexibility and variability?—needs to be asked in terms of conflicts over conditions defined both as use values and as exchange values.

Examples of capitalist accumulation impairing or destroying capital's own conditions, and hence threatening its own profits and capacity to produce and accumulate more capital, are many and varied. The warming of the atmosphere will inevitably destroy people, places, and profits, not to speak of other species life. Acid rain destroys forests and lakes, buildings, and profits alike. Salinization of water tables, toxic wastes, and soil erosion impair nature and profitability. The pesticide treadmill destroys profits as well as nature. Urban capital running on an "urban renewal treadmill" impairs its own conditions and hence profits, for example, through congestion costs and high rents.[22] The decrepit state of the physical infrastructure in the United States may also be mentioned in this connection. There is also an "education treadmill," a "welfare treadmill," a "technological fix treadmill," a "health care treadmill," and many others.[23] This line of thinking also applies to the "personal conditions of production . . . labor power" in connection with capital's destruction of traditionalist family life as well as the introduction of work relations that impair coping skills, and the presently toxic social environment generally. In these ways, we can safely introduce "scarcity" into the theory of economic crisis in a Marxist, not a neo-Malthusian, way. We can also introduce the possibility of capital *underproduction* once we add up the rising costs of reproducing the conditions of production. Examples include the health care costs necessitated by capitalist work and family relations; the drug and drug rehabilitation costs; the vast sums expended as a result of the deterioration of the social environment (e.g., the police and divorce costs); the enormous revenues expended to prevent further environmental destruction and to cleanup or repair the legacy of ecological destruction from the past; monies required to invent, develop, and produce synthetics and "natural" substitutes as means and objects of production and consumption; the huge sums required to pay off oil sheiks and energy companies (e.g., ground rent and monopoly profit); the garbage disposal costs; the extra costs of congested urban space; the costs falling on governments and peasants and workers in the Third World as a result of the twin crises of ecology and development; and so on. No one has estimated the total revenues required to compensate for impaired or lost production conditions and/or to restore these conditions and develop substitutes. It is conceivable that total revenues allocated to protecting or restoring production conditions may amount to one-half or more of the total social product—all unproductive expenses from the standpoint of self-expanding capital. Is it possible to link these unproductive expenditures (and those anticipated in the future) to the vast credit and debt system in the world today? To the growth of fictitious capital? To the fiscal crisis of the state? To the internationalization of production? The traditional Marxist theory of crisis interprets credit/debt structures as the result of capital overproduction. An ecological Marxist approach might interpret the same phenomena as the result of capital underproduction and the unproductive use of the capital produced. Do these tendencies reinforce or offset one another? Without prejudging the answer, the question clearly needs to be on the agenda of Marxist theory.

TOWARD AN ECOLOGICAL MARXIST ACCOUNT OF CAPITALISM AS A CRISIS-RIDDEN AND CRISIS-DEPENDENT SYSTEM AND THE TRANSITION TO SOCIALISM

Neither Marx nor any Marxists have developed a theory of the relationship between crisis-induced changes in the conditions of production and the establishment of the conditions of ecological socialism. In traditional Marxism, crisis-induced changes in productive forces and relations are determined by the need to cut costs, intensify labor, restructure capital, and so on. Forces and relations are transformed into more transparently social forms. In ecological Marxism, like traditional Marxism, capitalism is also not only crisis-ridden but also crisis-dependent. Crisis-induced changes in production conditions (whether crisis itself originates in capital overproduction or underproduction) are also determined by the need to cut costs, reduce ground rent, increase flexibility, and so on, and to restructure conditions themselves, for example, to expand preventative health measures, reforestation, reorganization of urban space, and the like.

There are two general, interdependent ways by which capital (helped by the state) changes its own conditions to weather crises and to resolve them in capital's favor. One is changes in conditions defined as productive forces. The other is changes in the social relations of reproduction of conditions. Changes in either typically presuppose or require new forms of cooperation between and within capitals and/or between capital and the state and/or within the state, or more social forms of the "regulation of the metabolism between humankind and nature" as well as the "metabolism" between the individual and the physical and social environment. More cooperation has the effect of making production conditions (already politicized) more transparently political, thereby subverting further the apparent "naturalness" of capital existence. The telos of crisis is thus to create the possibility of imagining more clearly a transition to socialism.

Crisis-induced changes in conditions as productive forces with the purpose of defending or restoring profit (exemplified by technological changes that lower congestion costs, increase flexibility in the utilization of raw materials, etc.) have the systemic effect of lowering the costs of reproducing the work force, making raw materials available more cheaply, and otherwise reducing cost and increasing flexibility. Whatever the immediate sources of the crisis, restructuring production conditions with the aim of raising profits is a foregone conclusion. More, crisis-induced changes in production conditions imply or presuppose more social forms of the social relations of reproduction of production conditions, for example, more direct forms of cooperation within the sphere of production conditions. An example of a change in production conditions today, and the associated change in the social relations of reproduction of production conditions, is integrated pest management which presupposes not only more coordination of farmers' efforts, but also more coordination of training and education programs.[24] Another example is preventative health technology in relation to AIDS and associated changes in community relations in a more cooperative direction.

The second form of restructuring is crisis-induced changes in the social relations of reproduction of production conditions introduced with the aim of exercising more control of production conditions, that is, more planning. Historically, planning has taken many forms, for example, urban and regional transportation and health planning, natural resource planning, and so on.[25] Whatever the immediate sources of crisis, the restructuring of these social relations with the aim of developing more control of production conditions is also a foregone conclusion. Furthermore, crisis-induced changes in the social relations of reproduction of production conditions imply or presuppose more social forms of production conditions defined as productive forces. An example of such a change today is "planning" to deal with urban smog which presupposes coalitions of associations and groups, that is, political cooperation, to legitimate tough yet cooperative smog- reduction measures.[26] Another example is the proposed restructuring of the U.S. Bureau of Reclamation which new technical changes in water policy presuppose.[27]

To sum up: Crisis forcibly causes capital and the state to exercise more control or planning over production conditions (as well as over the production and circulation of capital itself). We can be almost certain that the first major crisis of the new system of global capitalism will be the occasion for a host of new international planning instruments. Crisis brings into being new forms of flexible planning and planned flexibility, which increases tensions between a more flexible capitalism and a more planned capitalism—more so than in the traditional Marxist account of the restructuring of production and circulation because of the key role of state (and, increasingly, international) bureaucracies in the provision of production conditions. Crisis forces capital and the state to confront their own basic contradictions, which are subsequently displaced to the political, ideological, and environmental spheres (twice removed from direct production and circulation) where there is introduced more social forms of production conditions defined both materially and socially, for example, the dominance of political bipartisanship in relation to urban redevelopment, educational reform, environmental planning, and other forms of provision of production conditions. However, it is clear that technology and power embody one another at the level of conditions (as well as that of production itself), and hence that new forms of political cooperation hold out only tenuous promises of socialism. Again, nothing can be said a priori about "socialist imminence" excepting at a high level of abstraction. The key point is that capitalism tends to self-destruct or subvert itself when it switches to more social forms of the provision of production conditions via politics and ideology. The premise of this argument (like the argument of the present interpretation of traditional Marxism) is that any given set of production condition technologies and work relations is consistent with more than one set of social relations of reproduction of these conditions, and that any given set of these social relations is consistent with more than one set of production condition technologies and work relations. The "fit" between social relations and forces of reproduction of production conditions is thus assumed to be quite loose and flexible. In the crisis (in which by definition the future is unknowable), there is a kind of two-sided struggle to fit new production conditions defined as forces of production, not new production conditions defined as production relations, and vice versa, into more social forms without, however, any "natural" tendency for capitalism to transform itself into socialism. Urban and regional planning mechanisms, for example, may be a step toward socialism under certain political conditions, but not others. They are certainly a step toward more social forms of the provision of production conditions, hence making socialism at least more imaginable. So, too, regional transportation networks and health care services and bioregional water distribution (for example) may or may not be a step toward socialism; but they are certainly a step toward more social forms of the provision of production conditions.

In modern world capitalism, the list of new social and political forms of reproduction of production conditions is endless. It seems highly significant, and also theoretically understated within Marxism, that the world crises today appear to result in more, and require many additional, social forms not only of productive forces and relations but also of production conditions, although the institutional and ideological aspects of these forms are confusing and often contradictory, and although these forms should not be regarded as irreversible (e.g., reprivatization, deregulation, etc.). Yet it is conceivable that we are engaging in a long process in which there occurs different yet *parallel* paths to socialism, hence that Marx was not so much wrong as he was half-right. It may be that the traditional process of "socialist construction" is giving way to a new process of "socialist reconstruction," the reconstruction of the relationship between human beings and production conditions, including the social environment—a kind of Preservation First! politics. It is at least plausible that in the First World socialist reconstruction will be seen as, first, desirable, and second, as necessary; in the old Second World as equally desirable and necessary; and in the Third World as, first, necessary, and second, desirable. It is even more plausible that atmospheric warming, acid rain, and pollution of the seas will make highly social forms of reconstruction of material and social life absolutely indispensable.

To elaborate somewhat, we know that the labor movement "pushed" capitalism into more social forms of productive forces and relations, for example, collective bargaining. Perhaps we can surmise that feminism, environmental movements, and other new social movements are "pushing" capital and state into more social forms of the reproduction of production conditions. Labor exploitation (the basis of Marxist crisis theory, or the "first contradiction of capitalism," traditionally defined) engendered a labor movement which during particular times and places turned itself into a "social barrier" to capital. Nature exploitation (including exploitation of human biology) engenders an environmental movement (environmentalism, public health movement, occupational health and safety movements, women's movement organized around the politics of the body, etc.) which may also constitute a "social barrier" to capital. In Nicaragua, in the 1980s, the combination of economic and ecological crisis and political dictatorship in the old regime engendered both a national liberation movement and extensive ecodevelopment planning.

Concrete analysis of concrete situations is required before anything sensible can be said politically about environmentalism defined in the broadest sense, as well as capital's short- and long-term prospects. For example, acid rain causes ecological and economic damage. The environmental movement demands cleanup and restoration of the environment and protection of nature. This may restore profits in the long run or reduce government cleanup expenses, which may or may not be congruent with the short-and middle-term needs of capital. Implied in any systematic program of politically regulated social environment are kinds of planning that protect capital against its worst excesses, yet that may or may not be congruent with capital's needs in particular conjunctures. One scenario is that "the destruction of the environment can lead to vast new industries designed to restore it. Imagine, lake dredging equipment, forest cleaning machines, land revitalizers, air restorers, acid rain combatants."[28] These kinds of super-tech solutions would be a huge drain on surplus value unless they lowered the reproduction cost of labor power, yet at the same time they would help to "solve" any realization problems arising from traditional capital overproduction. Vast sums of credit money would be required to restore or rebuild the social environment, however, which would displace the contradiction into the financial and fiscal spheres in more or less the same ways that the traditional contradiction between production and circulation of capital is displaced into the financial and fiscal spheres today.

This kind of technology-led restructuring of production conditions (including technique-led restructuring of the conditions of supply of labor power) may or may not be functional for capital as a whole or for individual capitals, in the short or the long run. The results would depend on other crisis-prevention and crisis-resolution measures, their exact conjuncture, and the way in which they articulate with the crisis of nature broadly defined. In the last analysis, the results would depend on the degree of unity and diversity in labor movements, environmental movements, solidarity movements, and so on. And these are political, ideological, and organizational questions.

In any event, crisis-induced changes in production conditions necessarily lead to more state controls, more planning within the bloc of large-scale capital, a more socially and politically administered or regulated capitalism, and hence a less naturelike capitalism, one in which changes in production conditions would need to be legitimated because they would be more politicized—and one in which capitalist reification would be less opaque. The combination of crisis-stricken capitals externalizing more costs, the reckless use of technology and nature for value realization in the sphere of circulation, and the like, must sooner or later lead to a "rebellion of nature," that is, to powerful social movements demanding an end to ecological exploitation. Especially in today's crisis, whatever its source, capital attempts to reduce production and circulation time, which typically has the effect of making environmental practices, health and safety practices, and so on worse. Hence capital restructuring may deepen, not resolve, ecological problems. Just as capital ruins its own markets (i.e., realized profits), the greater is the production of surplus value, so does capital ruin its own produced profits (i.e., raise costs and reduce capital flexibility), the greater is the production of surplus value based on the destructive appropriation of nature broadly defined. And just as overproduction crises imply a restructuring of both productive forces and relations, so do underproduction crises imply a restructuring of production conditions. And just as restructuring of productive forces implies more social forms of production relations and vice versa, so does restructuring of production conditions imply a twofold effect: more social forms of production conditions defined as productive forces, and more social forms of the social relationships in which production conditions are reproduced. In sum, more social forms of production relations, productive forces, and conditions of production together contain with them possibilities of socialist forms. These are, in effect, crisis-induced not only by the traditional contradiction between forces and relations, but also by the contradiction between forces/relations and their conditions. Two, not one, contradictions and crises are thus inherent in capitalism; two, not one, sets of crisis-induced reorganizations and restructurings in the direction of more social forms are also inherent in capitalism.

CONCLUSION

Some reference needs to be made to post-Marxist thought and its objects of study, "postindustrial society," "alternative movements" or "new social movements," and "radical democracy."[29] This is so because post-Marxism has practically monopolized discussions of what Marx called "conditions of production." No longer is the working class seen as the privileged agent of historical transformation nor is the struggle for socialism first on the historical agenda. Instead, there is the fight for "radical democracy" by "new social movements" in a "postindustrial society."

These basic post-Marxist postulates deserve scrutiny, especially given post-Marxist readings of Marx and Marxism, and the political implications therein.[30] So does the declaration by many radical bourgeois feminists, ecofeminists, deep ecologists, libertarian ecologists, communitarians, and others. that Marxism is dead. In the present discussion, however, it is possible only to point out that in ecological Marxist theory, the struggle over production conditions has redefined and broadened the class struggle beyond any self-recognition as such, at least until now. This means that capitalist threats to the reproduction of production conditions are not only threats to profits and accumulation, but also to the viability of the social and "natural" environment as *means of life*. The struggle between capital and "new social movements" in which the most basic concepts of "cost" and "efficiency" are contended has two basic "moments." The first is the popular and nearly universal struggle to *protect* the conditions of production, or means of life, from further destruction resulting from capital's own inherent recklessness and excesses. This includes needs and demands for the reduction of risks in all forms. This struggle pertains to the form in which "nature" is appropriated, as means of reproduction of capital versus means of reproduction of civil society and human and other species life. The second is the struggle over the programs and policies of capital and state to *restructure* the production conditions, that is, struggles over the form and content of changes in conditions. New social struggles are confronted with both the impairment and also the crisis-induced restructuring of production conditions at the same time. Both "moments" of struggle occur outside the state and also within and against the state. Seen this way, the demand for radical democracy is the demand to democratize the provision and reconstruction of production conditions, which in the last analysis is the demand to democratize the state, that is, the administration of the division of social labor.[31] In truth, in the absence of struggles to democratize the state, it is difficult to take the demand for "radical democracy" seriously.

In post-Marxist thought great stress is placed on "site-specificity" and the "integrity" of the individual's body, a particular meadow or species life, a specific urban place, and so on.[32] The word "difference" has become post-Marxism's mantra, which, it is thought, expels the word "unity," which in the post-Marxist mind is often another way to spell "totalitarian." In well-thought-out versions of post-Marxist thought, the "site-specificity" which new social movements base themselves on are considered to make any *universal* demands impossible,[33] at least any universal demand beyond the demand for the universal recognition of site-specificity. This is contrasted with the bourgeois revolution which universalized the demand for rights against privilege, and the old working-class struggle which universalized the demand for public property in the means of production against capitalist property. However, our discussion of production conditions and the contradictions therein reveals clearly that there is a universal demand implicit or latent in new social struggles, namely, the demand to democratize the state (which regulates the provision of production conditions), as well as the family, local community, and so on. In fact, no way exists for diverse social struggles defending the integrity of particular sites to universalize themselves, hence win, and, at the same time, retain their diversity excepting through struggles for the democratic state and also by uniting with the labor movement, recognizing what we have in common, cooperative labor, thereby theorizing the unity of social labor.[34]

Moreover, post-Marxism, influenced by the "free rider problem" and problems of "rational choice" and "social choice" (all problems that presuppose methodological individualism), states or implies that struggles over production conditions are different than traditional wage, hours, and working conditions struggles because conditions of production are to a large degree "commons," clean air being an obvious example, urban space and educational facilities being somewhat less obvious ones. The argument is that struggles against air pollution (or capitalist urban renewal or racist tracking in the schools) do not have an immediate "payoff" for the individual involved, hence (in Offe's account) the phenomenon of cycles of social passivity and outrage owing to the impossibility of combining individual and collective action around goals that "pay off" for both the individual and the group. Again, this is not the place for a developed critique of this view, one that would begin with an account of how the process of social struggle itself changes self-definitions of "individuality." It needs to be said, however, that labor unions, if they are anything, are disciplinary mechanisms against "free riders" (e.g., individual workers who try to offer their labor power at less than the union wage are the object of discipline and punishment by the union). Further, it should be said that the "free rider" problem exists in struggles to protect the "commons" only insofar as these struggles are only ends in and of themselves, not also means to the specifically political and hence universal end of establishing a democratic state.

Also in relation to the problem of the "commons," and beyond the problem of the relation *between* the individual and the group, there is the problem of the relationship between groups and classes. Specifically, the struggles of "new social movements" over conditions of production are generally regarded in the self-defined post–Marxist universe as non–class issues or as multiclass issues. "Transformative processes that no doubt go on in our societies are very likely not class conflicts . . . but non–class issues."[35] Especially in struggles over production conditions (compared with production itself), it is understandable that these appear as non–class issues, and that agents define themselves as non–class actors. This is so not only because the issues (e.g., urban renewal, clean air, etc.) cut across class lines, but also because of the site-specificity and "people specificity" of the struggles, that is, because the fight is to determine what kind of use values production conditions will in fact be. But, of course, there is a class dimension to all struggles over conditions, for example, tracking in the schools, urban renewal as "people removal," toxic waste dumps in oppressed minority and poor districts and communities, the worker as the "canary" in the workplace, the inability of the unemployed and most workers to conveniently access "wilderness areas," and so on. Most problems of the natural and social environments are bigger problems from the standpoint of the poor, including minorities and the working poor, than for the salariat and the well-to-do. In other words, issues pertaining to production conditions are class issues (even though they are also *more* than class issues), which becomes immediately obvious when we ask who opposes popular struggles around conditions? The answer is, typically, capital, which fights against massive public health programs, emancipatory education, controls on investments to protect nature, adequate expenditures on child care, and demands for autonomy or substantive participation in the planning and organization of social life. What "new social movements" and their demands does capital support? Few, if any. What "new social movements" does labor oppose? Certainly, those that threaten ideologies of male supremacy and/or white race supremacy, in many instances, as well as those that threaten wages and jobs, even some that benefit labor, for example, clean air. Hence, the struggle over conditions is not only a class struggle, but a struggle against such ideologies and their practices. This is why it can be said that struggles over conditions of production (conditions of life and life itself) are not *less* but *more* than class issues. And that to the degree that this is true, the struggle for "radical democracy" is that much more a struggle to democratize the state, a struggle for democracy within state agencies charged with regulating the provision of production conditions. In the absence of such perspective and vision, "new social movements" will remain at the level of local and single issue struggles, which are bound to self-destruct themselves in the course of their attempts to "deconstruct" Marxism.

ACKNOWLEDGMENTS

The author wishes to thank Carlo Carboni, John Ely, Danny Faber, Bob Marotto, and David Peerla for their encouragement and helpful comments.

NOTES

1. Karl Polanyi, *The Great Transformation* (Boston: Beacon Press, 1957). Polanyi's focus was on capitalist markets, not the exploitation of labor.

2. World Commission on Environment and Development, *Our Common Future* (New York: Oxford University Press, 1987).

3. The closest to a true "Marxist" account of the problem is Alan Schnaiberg's *The Environment: From Surplus to Scarcity* (New York: Oxford University Press, 1980). This is a pathbreaking and useful work. The relation between the capitalization of nature and political conflict between states is another, and closely related, question (see Lloyd Timberlake and Jon Tinker, "The Environmental Origin of Political Conflict," *Socialist Review*, 15(6), November–December, 1985).

4. In the case of bad harvests, "the *value of the raw material . . . rises;* its *volume* decreases. . . . More must be expended on *raw material*, less remains for *labour*, and it is not possible to absorb the same quantity of labour as before. Firstly this is *physically impossible. . . . Secondly*, it is impossible because a greater *portion of the value of the product* has to be converted into raw material. . . . Reproduction cannot

be *repeated* on the same scale. A part of *fixed capital* stands idle and a part of the workers is thrown out into the streets. The *rate of profit* falls because the value of constant capital has risen as against that of variable capital. . . . The fixed charges—interest, rent—which were based on the anticipation of a *constant* rate of profit and exploitation of labour, remain the same and in part *can not be paid*. Hence *crisis*. . . . there is a *rise in the price of the product*. If this product enters into the other spheres of reproduction as a means of production, the rise in its price will result in the same disturbance in *reproduction* in these spheres" (Karl Marx, *Theories of Surplus Value*, [Moscow: Progress Publishers, 1968], Part 2, 515–516).

5. Apart from the degree of development, greater or less, in the form of social production, the productiveness of labour is fettered by physical conditions" (*Capital* [New York: Modern Library], 1:xxx). In *Theories of Surplus Value*, Part 3, p. 449, Marx states that the precondition for the existence of absolute surplus value is the "natural fertility of the land.

6. Michael Lebowitz, "The General and the Specific in Marx's Theory of Crisis," *Studies in Political Economy, 7*, Winter, 1982. Lebowitz includes as "general" barriers the supply of labor and the availability of land and natural resources. However, he does not distinguish between the supply of labor per se and the supply of *disciplined* wage labor. As for natural resources, he does not distinguish between "natural" shortages and the shortages capital creates for itself in the process of capitalizing nature or those created *politically by ecology movements*.

7. Karl Marx, *Capital* (Moscow: Foreign Languages Publishing House, 1962), 3:119, 792.

8. We can therefore distinguish two kinds of scarcity: first, scarcity arising from economic crisis based on traditional capital overproduction, that is, a purely social scarcity; and second, scarcity arising from economic crisis based on capitalistically produced scarcity of nature or production conditions generally. Both types of scarcity are ultimately attributable to capitalist production relations. The second type, however, is not due to "bad harvests," for example, but to capitalistically produced "bad harvests" as a result of mining, not farming, land; polluting water tables; and so on. This is an example of the "second contradiction."

9. There are two reasons why Marx ran from any theory of capitalism and socialism that privileged any aspect of social reproduction besides the contradiction between production and circulation of capital. One is his opposition to any theory that might "naturalize," hence reify, the economic contradictions of capital. His polemics against Malthus and especially his rejection of any and all naturalistic explanations of social phenomena led him away from "putting two and two together." Second, it would have been difficult in the third quarter of the nineteenth century to argue plausibly that the impairment of the conditions of production and social struggles therein are self-imposed barriers of capital because historical nature was not capitalized to the degree that it is today, that is, the historical conditions of the reproduction of the conditions of production today make an "ecological Marxism" possible.

10. State-of-the-art accounts of the problematic categories of productive forces and production relations include: Derek Sayer, *The Violence of Abstraction: The Analytical Foundations of Historical Materialism* (Oxford: Basil Blackwell, 1987),

and Robert Marotto, "Forces and Relations of Production" (Ph.D. dissertation, University of California, Santa Cruz, 1984).

11. To my knowledge, the phrase "ecological Marxism" was first coined by Ben Agger (*Western Marxism: An Introduction: Classical and Contemporary Sources* [Santa Monica, Calif.: Goodyear, 1987], 316–339). Agger's focus is "consumption," not "production." His thesis is that ever-expanding consumption required to maintain economic and social stability impairs the environment, and that ecological crisis has replaced economic crisis as the main problem of capitalism. This article may be regarded as, among other things, a critique of Agger's often insightful views.

12. According to Carlo Carboni, who also uses the expression "social reproductive conditions." I use "conditions of production" because I want to reconstruct the problem using Marx's own terminology and also because I want to limit my discussion mainly to crisis tendencies in the process of the production and circulation of capital, rather than to the process of social reproduction, that is, the reproduction of the social formation as a whole. This means that I will follow Marx's lead and interpret "production conditions" in "objective" terms, except in the last section.

13. External physical conditions include "natural wealth in means of subsistence" and "natural wealth in the instruments of labour" (*Capital*, 1:562).

14. Karl Marx, *Grundrisse* (Harmondsworth, U.K.: Penguin Books, 1973), 533. See also Marino Folin, "Public Enterprise, Public Works, Social Fixed Capital: Capitalist Production of the 'Communal, General Conditions of Social Production,' " *International Journal of Urban and Regional Research,* 3(3), September, 1979.

15. In a conversation with David Harvey, who pioneered the theory of the spatial configurations and barriers to capital (*Limits to Capital* [Oxford: Basil Blackwell, 1982]), tentative "permission" was granted the author to interpret urban and other forms of space as a "production condition."

16. The following is a deliberate "Smithian" simplification of the traditionally defined economic contradiction of capitalism which neglects Marx's critique of Smith, namely, that it is the rising organic composition of capital, not a falling rate of exploitation, that causes the profit rate to fall, even though capitalism "presents itself" otherwise. To be absolutely clear, the following account is not meant to review Marx's critique of capital fetishism or of Adam Smith et al. I put the contradiction of capitalism in its simplest terms with the twofold aim of (a) preparing a discussion of crisis-induced restructuring of the productive forces and production relations and (b) setting up a standard by which we can compare the "traditional" with the "nontraditional" or "second" contradiction of capitalism based on the process of capitalist-created scarcities of external and human nature.

17. "Cooperation" (e.g., "work relations") is both a productive force and a production relationship, that is, ambiguously determined by "culture," "technological necessity," and "power."

18. David Knight, *The Age of Science* (Oxford: Basil Blackwell, 1986).

19. This kind of formulation of the problem avoids the functionalism of the

"state derivation school" of Marxism as well as Weberian theories of the state that are not grounded in material relations and existence.

20. So-called external barriers *may be* interpreted as internal barriers, if we assume that (a) external nature is completely commodified or capitalized and (b) that new social struggles organized under the sign of "ecology" or "environmentalism" have their roots in the class structure and relations of modern capitalism, for example, the rise of the new middle class or salariat, which is the backbone of environmentalism in the United States.

21. "External and universal nature can be considered to be differences within a unity from the standpoint of capital accumulation and state actions necessary to assure that capital can accumulate. Yet the difference is no less significant than the unity from the standpoint of social and ecological action and political conflict. The reason is that labor power is a subject which struggles over health and the (natural) conditions of social health broadly defined, whereas the 'natural elements entering into constant and variable capital' are objects of struggle" (Robert Marotto, in a letter to the author).

22. "Economists and business leaders say that urban areas in California are facing such serious traffic congestion that the state's economic vitality is in jeopardy" (*New York Times*, April 5, 1988).

23. "If schools cannot figure out how to do a better job of educating these growing populations and turn them into productive workers and citizens, then the stability of the economy could be threatened" (Edward B. Fiske, "US Business Turns Attention to Workers of the Future," *International Herald Tribune*, February 20–21, 1988). Fiske is referring to minorities which today make up 17 percent of the population, a figure expected to jump to double by 2020. In the United States, health care costs as a percentage of GNP were about 6 percent in 1965; in 2000 they are expected to be 15 percent. "Health care has become an economic cancer in this country" (*San Francisco Chronicle*, March 14, 1988).

24. The well-known IPM program in Indonesia reportedly increases profits by reducing costs and also increasing yields. It depends on new training and education programs, coordination of farm planning, and the like (Sandra Postel, "Indonesia Steps off the Pesticide Treadmill," *World Watch*, January–February, 1988, p. 4).

25. For example, West German organized industry and industry–state coordination successfully internalizes many externalities or social costs. This occurs without serious harm to profits because the FRG produces such high-quality and desirable goods for the world market that costs of protecting or restoring production conditions can be absorbed while industry remains competitive (author's conversation with Claus Offe).

26. Christopher J. Daggett, "Smog, More Smog, and Still More Smog," *New York Times*, January 23, 1988.

27. The idea that crisis induced by inadequate conditions of production results in more social forms of production and production relations is not new in non-Marxist circles. Schnaiberg, in *The Environment*, linked rapid economic expansion to increased exploitation of resources and growing environmental problems, which in turn posed restrictions on economic growth, hence making some kind

of planning of resource use, pollution levels, and the like, essential. He interpreted environmental legislation and control policies of the 1970s as the start of environmental planning.

The idea that crisis induced by unfavorable production conditions (i.e., cost) results in more social productive forces, as well as production relationships, can be found in embryonic form in works such as: R. G. Wilkinson's *Poverty and Progress: An Ecological Perspective on Economic Development* (New York: Praeger, 1973), which argues that epoch-making technological changes have often resulted from ecological scarcities; and O. Sunkel and J. Leal's, "Economics and Environment in a Developmental Perspective" (*International Social Science Journal, 109,* 1986, 413), which argues that depletion of resources and scarcity increases the costs of economic growth because of declines in natural productivity of resources, and thus that new energy resources and technological subsidies (implying more planning) are needed.

28. Correspondence with Saul Landau.

29. The most sophisticated post-Marxist text is Ernesto Laclau and Chantal Mouffe's *Hegemony and Socialist Strategy: Towards a Radical Democratic Politics* (London: Verso, 1985). A home-grown example is Michael Albert et al., *Liberating Theory* (Boston: South End Press, 1986).

30. For example, Laclau and Mouffe's discussion of what they call Marxist "essentialism" violates both the spirit and the substance of Marx's theory of capital.

31. James O'Connor, "The Democratic Movement in the United States," *Kapitalistate, 7,* 1978. It should be noted that despite my familiarity with post-Marxist literature, I have been unable to find any reference to the division of *social* labor, so obsessed are the "theorists" with the division of industrial labor and the division of labor within the family. This absence or silence suggests that post-Marxism is at least in part recycled anarchism, populist-anarchism, communitarianism, libertarianism, and the like.

32. Accordingly to Carlo Carboni, "The challenge of specificity is propelled by all new social actors in advanced capitalist societies. It is an outcome of the complex network of policies, planning, and so on which are implemented by both capital and the state in order to integrate people while changing production conditions. On the one hand, this specificity (difference) represents the breakage of collective and class solidarity. On the other hand, it reveals both new micro-webs of social solidarity and the universalistic network of solidarity based on social citizenship" (letter to the author).

33. This and the following point were made by Claus Offe in conversation with the author, who is grateful for the chance to discuss these issues with someone who gracefully and in a spirit of scientific collaboration presents a post-Marxist point of view.

34. The issue in dispute is the post-Marxist claim that we have multiple social identities against the present claim that there exists a theoretical unity in these identities in the unity of the conditions of production and capital production and realization. On the level of appearances, it is true that we have multiple identities, but in essence the unity of our identity stems from capitalism as a mode of production. The trick is to make the theoretical unity a reality. An environmental

struggle may be an unintentional barrier to capital in the realm of accumulation while not being ideologically anti-capitalist. The question is how to make environmentalists conscious of the fact that they are making the reproduction of the conditions of production more social. The post-Marxists do not want to find a unity in the fragmented social identities we have. But even to build alliances between social movements some unity must be constructed. In the absence of an agreed upon telos of struggle, or any common definitions, dialogue cannot take place. If we are unable to agree on any terms and objects of struggle in what sense can we say new social movements are over what socialism means but in some sense we are required to struggle for a common language which will necessarily obscure particular differences. As capitalism abstracts out the social nature of labor in the exchange of commodities, it obscures what we have in common, cooperative labor, thereby fragmenting our identity. What is disturbing is the lack of any move on the part of the post-Marxists to theorize the unity of social labor" (letter to the author from David Peerla).

35. Claus Offe, "Panel Discussion," *Scandanavian Political Studies, 10,*(3), 1987, 234.

The Ecological Crisis:
A Second Contradiction of Capitalism?

VICTOR M. TOLEDO

Today, humankind is facing a planetary-scale ecological crisis, which it seems will reach its highest point in two or three decades. The ital-biosphere, or global ecological system, which encompasses all planet life and the space inhabited by it, is now under siege, threatened by human actions. This situation, without precedent in human history, seems to be the consequence of two ongoing main processes: the complete global integration of all human societies through communications, transport, technology, and economic trade; and the total colonization of the earth's spaces as a result of human population growth and expansion. Can we explain this global ecological crisis from a Marxist perspective? If so, which elements of the Marxist theory can be used as a coherent framework to develop an appropriate theoretical explanation?

James O'Connor proposes we attempt it through a basic economical approach: "The point of departure of 'ecological Marxism' is the contradiction between capitalist production relations and productive forces and conditions of production."[1] A similar idea was proposed by Skirbekk almost two decades ago.[2] According to O'Connor, *conditions of production* "are things that are not produced as commodities in accordance with the laws of the market (law of value) but that are treated as if they are commodities. There are three conditions of production: first, human labor power, or what Marx called the 'personal conditions of production'; second, environment, or what Marx called 'natural or external conditions of production'; third, urban infrastructure and space, or what he called 'general, communal conditions of production' "[3] He calls this the "second contradiction of capitalism" since it has as its basic cause "the capitalist self-destructive appropriation and use of labor power, space, and external nature or environment."[4]

O'Connor's proposition leaves me in a sea of doubts: Is the ecological crisis solely a consequence of an economic contradiction or, on the contrary, does it emerge from a highly complex set of causes: technology, demography, geography, culture, ideology, and forms of property? Can we accept the ecological crisis as proof of the self-destructiveness of capitalism? Are we facing a mere crisis of the economic system or a crisis of civilization (which implies a challenge not to an economic rationality, but rather to a "mode of life")?[5] Frankly, it is very difficult to qualify a proposition so general (and abstract) as O'Connor's idea of a "second contradiction." Although the environmentally destructive character of a market-oriented economy is more or less evident, there is, at hand, a great deal of evidence that refutes O'Connor's general hypothesis about the ecological guilt of capitalism. For instance, industrialization made whole towns and surrounding areas practically uninhabitable as long as a 150 years ago, and there is evidence of the depletion of natural resources in ancient civilizations such as Greece and Rome.[6] The problems of pollution, energy, and the destructive use of natural resources are also present, in the same proportion, in most of the countries of the ex-socialist block. There are, finally, ecologically successful (short-term?) experiences in market-oriented economies, and also environmental deterioration caused by demographic changes. On the other hand, the so-called ecological crisis includes a myriad of different phenomena. There are at least ten different processes provoking global environmental conflicts: deforestation, soil depletion, desertification, ocean and freshwater pollution, loss of biodiversity, toxic wastes, urban contamination, destruction of marine resources, the greenhouse effect, energetic misspending, and destruction of the ozone layer. Thus, how is it possible to attribute to capitalism the responsibility for every environmental problem?

I am afraid that while trying to relate Marxism and ecology, we a priori impute every recognized ecological problem to capitalism, creating commonplaces, not theory, and perpetuating a black tradition of dogmatism. In that sense, O'Connor's second contradiction could be useful as a general working hypothesis to be tested in future research, but not as a theoretical assumption. Is it, for instance, valid to put such ingredients as urban environment, health, forests, transportation, energy, or pollution together? I think O'Connor's provocative proposition should encourage an intensive and deep review of what could be called the "ecological thought" of Marx, which is yet an unexplored source of inspiration for the development of theories to explain the current environmental crisis and to offer political principles to the green movements. I understand this enormous challenge must be conceived as an original exploration of all that the Marxist tradition (and Marx) can offer to the comprehension of these new phenomena. The task should imply the review of an already important set of earlier works.[7] Needless to say, this challenge implies the review, with a new vision, of many of Marx's own texts. For instance, Parsons compiled over a hundred pages of selections from the writings of Marx (and Engels) on ecology.[8] Essential to this perspective is the review of the concept of nature in Marx, philosophical research into which was brilliantly initiated by Alfred Schmidt three decades ago.[9] Schmidt's book showed how Marx's thought, rooted in the nineteenth-century naturalist tradition, took as the theoretical keystone of his economic theory the basic principles of metabolism (*Stoffwechsel*) between nature and society. By doing that, Marx considered the human work process as an expression of the more general, eternal, presocial phenomenon of material exchange between human beings and the earth.[10] He also viewed the course of history as an increasing separation or conflict between nature and society (reaching its peak with capitalism), and the desirable future as its resolution.[11] In conclusion, while showing skepticism (or at least a cautious position) to theoretical generalizations such as those proposed by O'Connor in his postulation of the second contradiction, I suggest exploring more intensively the connection between Marxism and ecology by reviewing the "ecological thought" of Marx (and especially his so-called utopian or idealistic propositions). Thus, we can perhaps contribute to the formation of a new political philosophy: a new revolutionary theory urgently needed by a growing worldwide green movement in order to justify its political actions.

NOTES

1. James O'Connor, "Capitalism, Nature, Socialism: A Theoretical Introduction," *CNS, 1,* Fall, 1988, 23; reprinted as Chapter 9, this volume.

2. G. Skirbekk, "Marxisme et ecologie," *Esprit, 440,* 1974, 643–652. See also the English translation in *CNS,* 5(4), 1994; reprinted as Chapter 6, this volume.

3. James O'Connor, "Is Sustainable Capitalism Possible?" in *Conference Papers* (Santa Cruz, Calif.: *CES/CNS* Pamphlet 1), 12.

4. James O'Connor, "The Second Contradiction of Capitalism: Causes and Consequences," in *Conference Papers* (Santa Cruz, Calif.: *CES/CNS* Pamphlet 1) 4.

5. The idea that the ecological crisis is a crisis of civilization has been accepted by various authors such as A. Toffler, E. Tiezzi, E. Laszlo, G. Gallopin, M. Berman, and M. Grinberg. In V. M. Toledo, "Modernity and Ecology," *CNS*, 4(4), 1993, I have examined in some detail this idea, and I have postulated six main principles of Western (or modern) civilization, which should be theoretically and politically challenged by green movements: centralization, specialization, inequality, homogenization (ecogeographic, biological, genetic, cultural, and behavioral), (industrial) depredation (of rural and natural spaces), and autocracy.

6. J. D. Hughes, *Ecology in Ancient Civilizations* (Albuquerque: University of New Mexico Press, 1975).

7. See S. Moscovici, "Le Marxisme et la question naturalle," *L'Homme et la Societe*, *13*, 1969, 59–109; Skirbekk, "Marxism"; E. Romoren and T. I. Romoren, "Marx und die Okologie," *Kurbusch*, *33*, 1974, 175–187; J. P. Lefevre, "Marx et la nature," *La Pensee*, *198*, 1978, 51–62; J. Juanes, *Historía y naturaleza en Marx y el Marxismo* (Mexico City: Universidad Autónoma de Sinaloa, 1980); and E. Leff, "Alfred Schmidt y el fin del humanismo naturalista," *Antopología y Marxismo, 3*, 1980 (also reproduced as the first chapter in his book, *Ecología y capital* [Mexico City: Universidad Nacional Autónoma de México, 1986]). See also the articles by M. Sacristan and J.-G. Vaillancourt, *CNS*, *9*, 1992; and E. Leff, "Marxism and the Environmental Question," in *CNS*, 4(1), 1993.

8. H. L. Parsons, ed., *Marx and Engels on Ecology* (Greenwood Press, 1977).

9. A. Schmidt, *The Concept of Nature in Marx* (London: New Left Books, 1971).

10. This is a theoretical assumption I made in order to postulate a basic difference between *ecological* and *economic* exchanges, and to develop an ecological–economic conceptual framework of rural productive process. See V. Toledo, "Intercambio ecologico e intercambio economico en el proceso productivo primario," in *Biosociologia y Articulacion de las Ciencias*, ed. E. Leff (Mexico City: UNAM, 1981), and "The Ecological Rationality of Peasant Production," in *Agroecology and Small-Farm Development*, ed. M. Altieri and S. Hecht (Boca Raton, Fla.: CRC Press, 1990).

11. In 1844 (*Paris Manuscripts*) Marx wrote: "Communism as a fully developed naturalism is humanism and as a fully developed humanism is naturalism. It is the *definitive* resolution of the antagonism between man and nature, and between man and man. It is the true solution of the conflict between existence and essence, between objectification and self-affirmation, between freedom and necessity, between individual and species."

Capitalism:
How Many Contradictions?

MICHAEL A. LEBOWITZ

As James O'Connor continues to remind us, capitalism is subject to crises not only because capital spontaneously proceeds as if its realization requirements are independent of the rate of exploitation, but also because it similarly appropriates the natural conditions of production without regard for their requirements for reproduction. Thus, he proposes, there are not only crises characterized by capital's tendency for overproduction of capital (manifested in inadequate demand) but also those that reflect its tendency to underproduce (manifested in rising costs). Two contradictions—and both face capital at this point.

There is in all this the critical recognition of "the second contradiction," the understanding that capitalism's destructive appropriation and use of labor power, space, and environment flows from its very nature. And yet, it is appropriate to ask, first, whether the symmetry that O'Connor finds in the two tendencies toward crisis is what should be stressed and, second, whether there are two separate contradictions of capitalism or, alternatively, two forms of one contradiction.

226

Consider the first issue. Whereas it may be argued that realization crises have their proximate cause in the unintended consequences of individual capitals all pursuing cost-cutting measures, is it appropriate to focus (as O'Connor does) on the efforts of individual capitals to "externalize costs on to conditions of production" as the basis for the liquidity crisis characteristic of the second contradiction? What is thereby emphasized in both cases is the destabilizing actions of individual capitals, that is, the contradiction between actions of individual capitals and the requirements of capital as a whole.

Yet, while a hypothetical (but chimerical) single capital might avoid a realization crisis insofar as it is sufficiently farsighted never to drive the rate of exploitation higher than its own planned consumption and investment expenditures warrant, the same thought experiment conducted with respect to the appropriation and use of the conditions of production does not preclude the emergence of the second contradiction.[1] That single capital may indeed deforest, pollute, congest, and destroy health and environment, that is, degrade the conditions of production without any regard for their restoration. There is no reason to assume that this single capital (or a state that "looks out for the interests of capital as a whole") would regulate itself in such a way to ensure the "right" balance of appropriation and reproduction.

And why? Simply because there is no inherent and necessary reason for that single capital to bear the costs (whether they are the costs of functioning under such impaired conditions or the costs of repair). As long as it is able to shift those costs, capital (except in some hypothetical last instance) can ignore them. In this respect, the focus on the unintended consequences of the actions of individual capitals is misplaced.

To be sure, to the extent that individual capitals externalize their costs on to the conditions of production, they do increase the costs for other capitals. Similarly, since the strength of some capitals must appear as the weakness of others, local observation will support the position that falling profits and liquidity crisis are the result of shortages (of land, air, space, appropriate labor power, etc.) stemming from excessive appropriation of the conditions of production. Indeed, individual states (acting upon behalf of local capitals) may seek means of regulating them to aid them in the battle of competition. Yet, to understand capital, we need to consider capital as a whole rather than its appearance in competition. All these phenomena characteristic of the competition of many capitals should not detract from our understanding of either the inherent destructiveness (not self-destructiveness!) of capital as a whole nor its tendency to evade the costs of its destruction of the conditions of production.

As Marx commented, capital "takes no account of the health and the length of life of the worker, unless society forces it to do so."[2] So it is with all the conditions of production. Just as it was only the struggle of industrial workers to satisfy their own needs (at its core, "the worker's own need for development") that checked capital's tendency to appropriate the whole of their lives, so also is it the struggle (and that alone) of the modern multifarious proletariat to satisfy its needs for adequate health, education, environment, and so on, which checks capital's tendency and shifts the costs of impairment to capital.[3]

Rather than the unintended consequences of the actions of individual capitals, at the core of "the second contradiction" is that it is of the very essence of capital to determine the nature and extent of production without regard for human needs. But, then, that is true as well with respect to the conditions underlying the "first contradiction." In the one case, there is the tendency to produce without regard for natural conditions; in the other, to produce without regard to social conditions. Rather than two contradictions, there is indeed only one—that between the needs of capital and the needs of human beings. It takes (at least) two separate forms, and these forms interact in significant ways. But understanding the unity of those two forms is an important step in mobilizing people to do away not with the anarchy of individual capitals but with capital as a whole.

NOTES

1. For a discussion of the destabilizing actions of individual capitals and such a thought experiment conducted with respect to realization crises, see Michael A. Lebowitz, "Analytical Marxism and the Marxian Theory of Crisis, *Cambridge Journal of Economics, 18,* 1994, 163–179.

2. Karl Marx, *Capital* (New York: Vintage, 1977), 1:381.

3. Ibid., 772. See the discussion in Michael A. Lebowitz, *Beyond Capital: Marx's Political Economy of the Working Class* (New York: St. Martin's Press, 1992).

The Contradictory Interaction of Capitalism and Nature

ANDRIANA VLACHOU

The appropriation of nature under capitalism is complex and contradictory and gives rise to serious ecological problems. Ecological problems are the outcome of many diverse natural, economic, political, and cultural processes that are taking place within capitalist society and that interact with each other. However, as O'Connor has pointed out,[1] bourgeois naturalism, neo-Malthusianism, Club of Rome technocratism, romantic deep ecologyism, and United Nations one-worldism—despite their interesting insights on different aspects of ecological problems—fail to address the *class dimensions* of these problems. Moreover, as these analyses both ignore class exploitation and effectively reduce the nature–society relationship to a single ultimate determinant, they cannot produce a satisfactory knowledge of how the interaction between nature and society is shaped, and thus changed, over time and across different societies.

I consider the class process[2]—the production, appropriation, and distribution of surplus labor—an important aspect of capitalist society that shapes the relationship between nature and society. In turn, natural

229

processes, along with other non–class processes, condition and bring into existence the class process. Thus, transformations wrought in nature by capitalism are articulated with the capitalist class process; this interaction can become a point of departure for a Marxist discourse on capitalism and nature.

However, before I comment on some of the economic aspects of capitalism's interaction with nature, I would like to emphasize that in no way do I embrace the economism of traditional Marxism.[3] I conceive of society as an overdetermined totality that cannot be reduced to a mere effect of economy or any other of society's constituent aspects. Moreover, for me, the class process itself comes into existence only out of the interaction of different natural, political, cultural, and other economic processes; thus, every aspect of society should be accounted for in class analyses.

A Marxist discourse on nature can preserve, I think, the *centrality of the class concept* since the interaction between society and nature is mediated by social labor, which is performed within class relations in class societies. Thus, as we experience nature in historically produced forms,[4] we can start producing a knowledge of capitalism's appropriation of nature by directly addressing, instead of ignoring, the class aspects of this process. Moreover, nonreductionist Marxism is well situated to theorize all the aspects of the relationship between nature and society, and we should encourage Marxists from different disciplines to produce such specific knowledges. I, as an economist, shall try to discuss some economic aspects of capitalism's interaction with nature.[5]

Capital needs natural conditions and resources to be available in requisite quantities and qualities. Land, minerals, water, clean air, and other natural conditions are conditions of existence for the capitalist class process; they constitute elements of constant capital and they also sustain life, and, for that matter, they secure the existence and the reproduction of the valuable commodity labor power. However, capitalist development itself is contradictory and uneven; thus, capital's expansion over time might threaten the very conditions of its existence, including its natural ones.

Environmental degradation, as experienced today, has been caused, among other factors, by the technical and economic modes in which nature is appropriated by the capitalist class process. Technology, an overdetermined social process, is closely associated with the extraction of relative surplus value and cost reductions. Many environmental resources were ignored in the shaping of capitalist development. These resources were common property, and did not command a market price. They were easily appropriated by capital in the form of waste receptacles. Pollution-abatement technologies (to the degree that they were available) raised costs, so that they were not introduced on the basis of the initiative of individual capitals. In this way, capitals tended to jeopardize their own existence as they exceeded the "carrying capacity" of nature, resulting in the destruction of natural processes and conditions.

Individual capitals may also deplete nature as a source of elements of constant and variable capital as they extract natural resources in a very shortsighted manner imposed on individual capitalists by virtue of the structure of capitalist competition pertaining to accumulation rates, technical change, size of capital, and so on. In their search for cheap raw materials and wage goods, individual capitalists first use high-quality or easily accessible natural resources without any consideration for their availability in the future. In this way, they succeed in keeping their cost of production low and maximizing their profits. However, since many strategic natural resources are reproduced by nature at a slow rate, they run up against "limits." They tend to exhaust first the high-quality or easily accessible reserves and then use other resources of lower quality, thereby giving rise to higher costs of production. This kind of "scarcity," however, is socially and historically produced; it is clearly related to the way that natural resources are appropriated by the capitalist class process.

Another kind of "scarcity" in the case of natural resources is created by the existence of monopolies in resource industries. Exclusive private or "state" property of certain strategic resources, whose reproduction by nature is not easy, restricts access to these resources and creates a natural monopoly. High rents might be charged to individual capitals or consumers who want to gain access to these resources. Individual capitals experience the increases in the monopoly prices of natural resource commodities that they buy as increases in costs, while workers with constant nominal wages find their real wages falling. The case of oil in the 1970s was an excellent example of changes in monopoly positions resulting in higher prices. At the time, however, a significant effort was made to convince the public that higher prices were related to an increased natural scarcity of oil. But "natural limits" in certain epochs have been the outcome of historical processes that shape society's relation to nature and cannot be defined only by reference to "natural conditions." This was especially true for the 1970s.[6]

The degradation of nature produced by capitalism, on the other hand, reacts back upon the class process as it threatens capital's conditions of existence and produces changes toward securing these natural conditions. However, these changes are shaped by class and non–class struggles that are fought at all the levels of society. Moreover, their outcome is uncertain as various changes might be conducive or inimical to further capitalist development.

These struggles contain political, ideological, economic, and other aspects. But I would like to emphasize some of the economic aspects. There is an intercapitalist struggle, that is, there is competition between individual capitals for their survival, which leads in the first instance to pollution or depletion of natural resources. In the absence of any regulation, if some capitals choose to control their emissions, they might find their economic position deteriorating, and hence risk being driven out of business. In the second instance, however, as some individual capitals experience increases in costs due to ecological degradation, they start fighting other capitals in an effort to make these capitals "internalize" environmental costs or reduce the monopoly prices they charge. The tourist industry, for example, struggles against industrial firms that use lakes, rivers, or the sea as sinks for their waste, causing damages to the former. Non–energy firms fought against oil companies in the 1970s to keep the price of oil low.

The state might also be called upon by different capitals to mediate access to nature. The state then becomes the site of these struggles. For example, the state can nationalize or regulate the energy industry; in this case it does not simply secure the political and ideological conditions for capitalist extraction but also the economic ones.

Ecological destruction has direct consequences on working people as well. People find out that the means of life are not so easily available any more. Pollution, for example, has significant negative effects on human health, resulting in various diseases and increased medical care costs. Certain wage goods or services, for example, heat, or houses in relatively clean areas, become more expensive. In other words, environmental degradation increases the cost of the reproduction of labor power. Consequently, capitalist firms also experience these developments in different ways. Absence from work due to health problems and declines in productivity are two forms. Or, people may start fighting at the point of production for wage increases to compensate for the cost increases in wage goods and services, or people may organize in local movements to fight for regulation against polluting firms in an effort to protect their conditions of life. All these struggles, no matter where they start, have corresponding class movements and shape the extraction process as well. In addition, all these struggles may be fought inside or outside the state. Social movements, state agencies, and production sites are all loci for the shaping of the society–nature relationship. The struggle, for example, over the establishment of the operation of a nuclear power station is fought at the local level, within and against the state, and at the production site.

Historically, these struggles gave rise to environmental regulation by the state, and "environmental" industries producing commodities or services that "protect" the environment or mitigate "scarcity," and, at the same time, make a profit just like any other business. An ecological ethic may propel not only social movements, but environmental business too. Pollution might be reduced and "scarcity" might be mitigated. In this sense, I think *there is no a priori tendency for capitalism to produce environmental crises.* It is possible for capitalism to develop new modes to secure its natural conditions of existence. However, as a result, capitalism might threaten other aspects which are important for its existence. For example, pollution abatement or environmental education might absorb a significant part of surplus value, which would otherwise be available for technical restructuring and the expansion of profits, not to mention other social or natural needs. As the capitalist appropriation of nature is complex and contradictory, as well as mediated by all the other processes of society, there is a real possibility that crises of capitalism can emerge out of ecological problems.[7]

It should also be noted that under capitalism measures to protect natural conditions cannot simply be considered to be the product of successful struggles on the part of the ecological movements. Many of them might be compatible with the long-term existence of capitalism. This point was also made by T. Kyprianidis as a critique of certain Greek ecologists who tend to ignore the *interactive nature of ecological problems as both causes and effects of the shaping and reshaping of capitalism today.*[8]

Nor are these measures the simple result of a state agency's effort to regulate capital's access to nature, especially when this agency is considered primarily as a political and ideological site. This line of thinking might avoid economism but it may end up with voluntarism as it privileges the state and its political and ideological aspects or, alternatively, non–state institutions, especially the social movements.

Environmental issues, and struggles over them, are as much political and cultural as they are natural and economic. Involved in them are social movements as well as capitalists and workers, and in many cases they occupy different and contradictory positions in these struggles. Thus, the outcomes become overdetermined.

Moreover, in socialism, new political and ideological processes need to be combined with alternative postcapitalist economic (productive and distributive) processes in order to create a new articulation between nature and society. Social movements and struggles can effectively pave the road to (and shape) the new society. But capitals are also in constant change and readjustment with the aim of meeting these challenges. This contradictory interaction, however, is *open-ended.* It is in this open-endedness that the significance of social movements and struggles is grounded, so that the transcendence of capitalism can be hoped for and worked for.

NOTES

1. James O'Connor, "Capitalism, Nature, Socialism: A Theoretical Introduction," *CNS, 1,* Fall, 1988, 13; reprinted as Chapter 9, this volume.

2. The theorization of class as a social process overdetermined by all other natural and social processes has been developed by Steven Resnick and Richard

Wolff in *Knowledge and Class: A Marxist Critique of Political Economy* (Chicago: University of Chicago Press, 1987).

3. I have developed an overdeterminist account of the relationship between society and nature in "Reflections on the Ecological Critiques and Reconstruction of Marxism" *Rethinking Marxism, 7,* Fall, 1994, 112–128. An overdeterminist standpoint does not render theorizing impossible or meaningless; on the contrary, it calls for complex social analyses. In this sense, to produce a knowledge of the interaction of capitalism and nature—and also to change this relationship—is to unfold the many ways in which social processes interact with natural ones, their effectivity in their mutual constitution, and the contradictions that are brought about by their complex interaction.

4. O'Connor, "Capitalism, Nature, Socialism," 13.

5. I have presented many of the arguments that follow in "A Marxist Analysis of Environmental and Natural Resource Problems," *Theses, 33,* October–November–December, 1990 (in Greek), 123–140.

6. For a further analysis, see A. Vlachou, "A Dynamic Analysis of Energy Demand in U.S. Manufacturing during the 1970s" (Ph.D. dissertation, University of Massachusetts, 1983).

7. Similar points were also made by Stephen Resnick and Richard Wolff in "Nature, Class, and Crisis" (Mimeo, University of Massachusetts, Amherst).

8. T. Kyprianidis, "Notes on Ecology and the Left," *Theses, 29,* October–December, 1989, 31–38.

The Second Contradiction in the Italian Experience

VALENTINO PARLATO
GIOVANNA RICOVERI

THE POTENTIAL
OF THE SECOND CONTRADICTION

The "second contradiction of capitalism" means (in our opinion) that there is a mortal, even a suicidal, conflict between capitalist society and nature. The true implications of this conflict, in our view, have not yet been fully perceived, because it represents something new, something that is not yet clearly understood. There is, however, one point that we do understand. In discussions of the second contradiction, there is the danger of representing it as a conflict between two "things," nature and industry, that is, to reify and hence to misconstrue the whole problem. Our first point is that "industry" should be defined in qualitative terms, to include people's habits, cultural relations, and the anthropological setting generally, and, similarly, that "nature" needs to be defined as movement and change, that is, also qualitatively.

Within "economistic" left and social democratic circles (although not within the green discourse), a limited or static, reified view of the relation between nature and industry is the dominant one. This may, in fact, reflect

235

the same limited view of the conflict between capital and labor so often found within these same circles—as purely economic, quantitative, nondevelopmental. Whatever the case, the fact is that the second contradiction did make its appearance in history, at this time, in quantitative terms or with a strong economic bias, so to speak. No conflict between capitalist society and nature appeared to exist so long as the latter was open to seemingly infinite exploitation, free of charge, a *res nullius*. In other words, this conflict came to the fore when the limits and costs of nature became obvious to (almost) all.

The political goal of economistic leftists, social democrats, and trade unionists who reduced the first contradiction of capitalism—between capital and labor—to a set of quantitative relations was to modify capitalist production relations in the direction of more fairness or equality, not to abolish the capital–wage labor relation per se. Not the end of wage labor, but improvements in wages, employment, and so on were the "telos" of these traditions. The danger for the Left (and the environmental movement) is to reduce the second contradiction—between capital and nature—to another set of quantitative relations, for example, "nature prices," green taxes, subsidies to nonpolluting energy sources, and the like. This is a danger because the domination and "exploitation" of nature has resulted not only in nature becoming too "costly," but also in the loss of control over our own lives, which necessarily includes our relationship with nature. Put another way, just as labor power is treated as if it is a commodity, so does "capitalist nature" mean that nature is commodified, hence alienated from us, and ultimately reified into a "thing." And just as the capital–labor relation should be seen in qualitative terms, that is, as loss of power, alienation, and so on, so should the capital–nature relation be interpreted as powerlessness, and alienation.

We think that this is, in fact, what is happening, that the transformation of nature into a commodity, into capital (called "natural capital" by bourgeois ecological economists), has become increasingly unacceptable from a human point of view, and that more people are aware of their estrangement from their natural conditions of existence—albeit in an overly passive way, politically speaking. The reason we believe this is that poor environmental conditions have brought about a decline in basic human values and have made life more barbaric (especially in the South but also in the North). These conditions have resulted in social disorientation, or a sense that people have lost track of their lives.

Now more than ever, we think, the liberation of women and men presupposes the liberation of nature and a halt to the commodification and capitalization of the environment. Given the "discovery" that we, too, are nature, there is also the fact, and a growing awareness of this fact, that the destruction of the environment means that we are losing our own sense of self and of life possibilities. Just as in previous times, for example, when wage workers (and people in general) refused to eat cornmeal without salt, there is a growing refusal to breath carbon monoxide, to live in a treeless landscape, or to die from skin cancer due to holes in the ozone layer.

THE PROBLEM OF RED–GREEN–FEMINIST UNITY

Given the economic and political conjuncture in Italy today, it is understandable that the slogan "red–green–feminist unity" seems to be overly simple, hence not very convincing. Certainly an analysis of this slogan is in order. First, the main consequence of the awareness of the second contradiction is the cultural and specifically political need to make alliances—always a strong political trend in the history of the worker's movement and the Left. Without wanting to exaggerate our point by the use of analogies, Lenin's concept of an alliance between workers and peasants—in a situation where the contradictions within the factories were very different than those on the land—is not unlike O'Connor's concept of a "red–green–feminist" alliance—under conditions in which contradictions between capital and labor, capital and nature, and men and women have different forms and political trajectories.

Another way to address the problem is this: Some of us lived through a period of rapid and seemingly limitless industrial growth, and also believed (on the basis of a simplified reading of Marx) that capitalist society would become more polarized. We thought that growing proletarianization would cancel out differences within the working class, broadly defined. And just as we are engaged in self-criticism for our neglect of the "problem of nature," so, too, are we questioning our old views on the "problem of labor." As differences return, or emerge, and also as our awareness of difference becomes more acute, we can see more clearly the problems inherent in the concept of a red–green–feminist alliance. These are precisely the differences between the working classes hit by the first contradiction; all of us hurt by the second contradiction; and women are assailed by the gender contradiction. Parallel although overlapping axes of problems and struggles exist. If we add to these struggles those of oppressed minorities and those of pacifists and others who are trying to address reemergent nationalism and war within the gates of Europe, it becomes clear that there is no present structure that unites these and related diverse, critical, social, and political strands. But it is also clear that there needs to be such a structure—of unity within difference and difference within unity—that is, that a red–green–feminist alliance is necessary and urgent. Put bluntly, no one group, class, or faction can possible gather the strength and will to confront global capital on the nature, labor, gender, place, and antiracist fronts simultaneously.

SUSTAINABLE DEVELOPMENT: RESISTANCE AND TRANSITION

With regard to the environmental question and the culture of the Left, what kind of development is humanly bearable is clearly a central question. From the standpoint of the problematic of *CNS*, we would like to suggest that the issue of what human beings can and cannot endure is linked to the second contradiction of capitalism. By contrast, reformism and the "Keynesian pact or accord" is related to the first contradiction. Is this a mistaken contrast or analogy, or a useful research suggestion? One way to answer this question is to suggest that—given that the environmental question has arisen because of the capitalist commodification of nature— "sustainable development" as defined up to the Rio 1992 conference is not a solution to the problem. The reason is that it is more difficult to regulate the second contradiction than the first—the former has created more rigidities than the latter. The good intentions of those who support "sustainable development" notwithstanding, the fact is that "sustainability" implies the total commercial exploitation of nature. Air, water, and countryside become merchandise, for which we pay more or less oligopolistic prices related to quantity, quality, and market conditions. Just as reformism and Keynesianism for a time successfully regulated capitalism, so, too, the path of sustainable development can become the driving force to give new life to a crisis-prone capitalist system, especially in "newly industrializing countries." (Consider, e.g., a world in which fresh air is likened to pensions for everyone.)

If this is true, it follows that the "ecological revolution"—the transformation of the relations we have with nature—is more important than the "proletarian revolution" as it was traditionally understood. We add that revolutions are gradual, as well as sudden. This means that there are new and urgent needs the solutions to which cannot be postponed to the "day after the revolution." (This problem has many analogies with those of the reformist working-class movement.) Hence, it is not only useful but necessary to discuss the concept of transition, and transition objectives, which pertain to defensive struggles as those aimed toward radical change. Very relevant to this point is the recent debate in *Il Manifesto,* titled "Work, Resistance, Transition, Planning."

This isn't a philosophical question. Rather, it leads back to the subject of a red–green–feminist alliance, that is, the formation of a bloc (perhaps in the Gramscian sense, perhaps in some new sense). This subject returns us to the need for a less schematic and simplified analysis of the second contradiction than the one that we have so far developed in relation to the first.

Summing up, the considerations introduced above seem to us to constitute the coordinates of the new cycle of social, political, and cultural conflict, one that will produce its own subjects, articulations, and organizational forms. Yet we need to remember that "everything is fresh off the machine" and that we are still at the beginning of a new phase.

PART IV

Critical Voices

INTRODUCTION
TO PART IV

There is a well-established and widely believed "green" critique of Marxism. In the United Kingdom, a leading green activist, Jonathan Porritt, referred to capitalism and communism as "tweedledum and tweedledee"[1]: both are supposed to be equally hostile to green hopes and values. In the United States, Murry Bookchin has been a leading critic of Marxism from the standpoint of his "social ecology." Marxism has been berated for its "productivism"; its reliance on scientific and technical advances (the famous "development of the forces of production") to deliver a future paradise of material abundance; and for its commitment to the modernist project of "mastery of nature."

This volume has, I hope, at least demonstrated that the situation is far more complex than this. Partly, this is because the legacy of classical Marxism is itself ambiguous or, more often, just undeveloped, in precisely those areas where the green critique is most pressing. More importantly, however, what the standard versions of the green critique of Marxism overlook is that Marxism is itself an intellectual and political tradition *with a history:* that is to say, like all living traditions it is engaged in a dialogue with other traditions, responding to new problems and experiences, and critically evaluating and revising its own heritage as it does so. None of the contributions in this book would have been written if this were not true.

The various attempts to reconstruct the materialist conception of history in the face of the challenge of ecological crisis and the new politics of the environment have, indeed, led many Marxists themselves to abandon beliefs and ways of thinking that others have seen as definitive of Marxism. So, are they still Marxists? This is a question to which a Stalin or a McCarthy would have demanded a straight answer. For the rest of us, it should not matter. The key point, for most of us, would be to play a part in developing a framework of ideas that will empower people in the struggle against an oppressive, exploitative, and destructive "civilization." Despite their differences of opinion, the contributors to this volume are united in thinking that the historical materialist tradition can make a distinctive and valuable contribution to that struggle, and that other movements would be poorer and weaker for the loss of the unique perspectives and insights historically associated with Marxism.

However—and, again, this would be widely agreed—for these insights to continue to play their part requires Marxists themselves to maintain a continuous dialogue both with their own past and with the ideas of other traditions. In Part IV of this volume, titled "Critical Voices," we offer three contributors who demonstrate the value of this continuing openness to critical challenges. These contributors give voice to diverse strands of feminist and ecocentric skepticism about our attempts to create a synthesis of green and socialist thought. However, what makes these contributions different from the more widely shared green critique of Marxism is that their writers have either themselves made significant contributions to the red–green project, or, at least, acknowledge the continuing process of critical self-renewal that has been going on among the "reds." What they offer, then, is thoughtful and constructive dialogue, not hasty, knee-jerk rejection.

This part of the collection begins with a powerful statement by Mary Mellor, to the effect that the synthesis of feminism and ecology, which has been happening independently of the green–Marxist encounter must be fully recognized. She acknowledges the relevance of two concerns that ecological Marxists might have about "ecofeminism": that the corrupting power of capitalism will be insufficiently recognized, and that "essentialist" accounts of women and nature will lead to "reification" and stultification of change. However, as Mellor points out, there are many feminisms, just as there are many socialisms: ecofeminism is not necessarily "essentialist" about either women or nature, whereas traditional Marxism *could* be seen as "essentialist" about production.

In Mellor's view, capitalist economic relations involve both the exploitation of wage labor and, through the public–private opposition, reproductive, procreative work. In their focus on production at the expense of reproduction, traditional Marxists (except perhaps Engels) have shared the patriarchal bias of capitalism itself. She recognizes that there are many women wage workers, and that men are not debarred by their biology from taking part in nurturing, caring, procreative activity outside the wage economy. Nevertheless, there is, as a matter of historical and sociological fact, a strong association between wage work and male experience, on the one hand, and nurturing, parenting, and the like as women's experience. She endorses the work of ecofeminists such as Vandana Shiva and Maria Mies who also extend this analysis to include the role of Third World women in the international division of labor, at the interface between fragile ecosystems and the direct meeting of physical and emotional needs.

According to Mellor's approach, Marxist analysis needs to take account of both feminist and ecological thinking by recognizing three interrelated material bases of society: the forces and relations of production, the forces and relations of reproduction, and the society–nature relationship. This, presumably, would strengthen the explanatory power of a materialist approach to historical understanding, but even this would be an insufficient response to the feminist case. Mellor is also concerned with questions of the *normative* basis for a feminist green socialism, and of the kind of agency that would carry this project forward. Drawing on the "feminist standpoint" theories of Nancy Hartstock and others, she argues that the characteristic—but *not* biologically determined—experiences of women in relation to both reproduction and ecological maintenance are the material basis for a set of values that must be *prioritized*, not just incorporated, in socialist practice.

Kate Soper's response, while it endorses the main outlines of Mellor's argument, is a further illustration of the diversity of feminist perspectives on this range of issues. Though Mellor takes great pains to defend insights from ecofeminism without making concessions to "essentialism," Soper is unclear about how far this has been achieved. On the key question of whether the aim should be to revalue the caring, nurturing and life-sustaining activities traditionally associated with women, or, rather, to transform the institutions that segregate and allocate these activities to women, Soper finds Mellor equivocal. Also, the complexities of such concepts as "nature" and "biology" (let alone "materialism" and "essentialism") are such that Mellor's argument leaves many questions unanswered.

Finally, this book concludes with two extracts from an important recent work by Robyn Eckersley. Eckersley, as an advocate of ecocentrism, remains critical of what she sees as the inherent anthropocentrism of the ontology and value perspective of Marxism. However, her critical relationship to Marxism is more sympathetic than this stance might suggest, since she, unlike many critics, acknowledges the extent to which Marxists have critically revised their own traditions to take account of ecological issues. And, on some important questions, Eckersley's brand of ecocentric political theory is on the same side of the debate as ecological Marxism.

Eckersley's book critically reviews the diversity of "ecopolitical thought" around what she sees as a basic cleavage between "ecocentric" and "anthropocentric" perspectives. For the latter, the nonhuman world is seen as a storehouse of resources, which are valued instrumentally, only insofar as they can serve human needs. By contrast, ecocentric philosophies see both humans and nonhumans as interconnected parts of a wider totality, in which nonhumans as well as humans are assigned intrinsic value, or moral worth. But viewed in terms of another distinction that Eckersley draws, both ecocentrics and anthropocentrics may appear on the same side as one another. The "survivalist" ecopolitical thinking that was at its height in the wake of *Limits to Growth* and "Blueprint for Survival," in the early 1970s, is contrasted with "emancipatory" ecopolitical thought. This latter group of perspectives takes the recognition that there are limits to growth as the starting point for a renewal of critical thinking about the ecologically disastrous view of "progress" that has so far prevailed. Theories of participatory democracy, an end to alienated relationships, and a shift from quantitative to qualitative concerns are fused with the new politics of ecology.

While the Marxist tradition and emancipatory ecocentric theorists both reject "survivalist" ecopolitical thinking, their emancipatory alternatives divide along anthropocentric and ecocentric lines. However, the situation is even more complicated than this, and for two reasons. The first reason is that Eckersley, along with many other writers on environmental issues, recognizes the great diversity of approaches represented within the broad category of "environmentalism." Although the division between the anthropocentric and the ecocentric approaches is central for her, she is also sensitive to the varieties of approach *within* what she calls anthropocentrism.

She distinguishes four main "streams" of environmentalism that fall short of ecocentrism: *resource conservation,* concerned primarily with efficient use of resources and avoidance of waste; *human welfare ecology,* which seeks to protect the human environment from degradation and recognizes a wider range of human psychological and social needs in relation to nature; *preservationism,* which seeks to protect areas of wilderness from "development"; and *animal liberation.* Only the first two "streams" of thought are unequivocally anthropocentric. The preservation of wilderness may be supported for aesthetic or spiritual reasons, but also for reasons independent of any human interest. Animal liberation does assign intrinsic value to nonhuman beings, but does so only on the basis of *similarity* to humans, and is also limited, from an ecocentric point of view, by its "atomistic" ontology.

The ecocentric approach Eckersley favors is not antagonistic to these other streams of environmentalism, but rather it comprehends and surpasses them: its holistic ontology, which *includes* humans within nature, enables ecocentrism to favor the flourishing of *both* humans and nonhuman nature. The widespread accusation that ecocentrism is misanthropic is directly refuted. So, the ecocentric objection to Marxism as inherently anthropocentric is not necessarily that it is pointed in the wrong direction, but, perhaps, that it has not (yet) gone far enough in the right one. There will be wide areas where ecocentric and anthropocentric proponents of human emancipation share common ground, but also areas where ecocentrics will go beyond anthropocentrics in their wider vision of the proper subjects of moral concern and emancipation. At the same time, of course, it has to be recognized that this wider vision of emancipation will also reflect back on what is seen as important for *human* emancipation.

A second reason why the ecocentric approach to Marxism is more complex than it appears to be at first sight has to do with the range of different social and political theories that claim the title "Marxism." In the first of the two extracts reprinted here, Eckersley acknowledges the existence of two main traditions: "humanist" and "orthodox" Marxism, relying primarily on the earlier and the later works of Marx, respectively. She recognizes that both traditions have addressed ecological questions, and she discusses each of them from her own ecocentric standpoint.

"Orthodox" eco-Marxism attributes environmental degradation to capitalist relations of production, and sees the socialist transcendence of capitalism as the solution to these ecological problems. However, the classical Marxist tendency to put a positive value on the development of science and technology, in line with the "modernist" project of ultimate mastery of nature, is not called into question. The principal *ecological* objection to capitalism is that it *limits* the development of sustainable mastery of nature, while the principal *social* objection is that the benefits of mastery are not shared by the whole community. For ecocentrics, the ecological argument is not taken far enough. In particular, orthodox ecological Marxism has no place for the value of nonhuman beings as unfolding or flourishing in their own ways. A consequence of this attitude is that the project of a more complete, albeit sustainable, domination of nature is not challenged by orthodox eco-Marxists, so the wider problems of instrumental rationality and modern industrial technology (the "forces" of production) remain unaddressed.

However, many Marxists, particularly in the humanist camp (Eckersley mentions Gorz, Marcuse, and Donald Lee most frequently) do recognize that not only capitalist relations, but also the forces of production they have engendered need to be transformed. For this vision of the future society these writers turn to the early works of Marx for inspiration. Donald Lee comes closest to what Eckersley regards as an ecocentric position, relying on Marx's concept of nature as humanity's "inorganic body," his view of alienation as including alienation from nature, and his vision of an emancipated future in which humans realize their "species being" through reconciliation with nature. But Marcuse and Gorz are more characteristic of humanist eco-Marxism, and truer to Marx, in that they make future human freedom dependent upon a technical mastery of the forces of nature. For Eckersley, the opposition between necessity and freedom, with its associated opposition between work and leisure, is central to Marx's thinking. It implies that humans can be free only to the extent that they subordinate nature and society to their purposes.

This inverse relationship between human freedom and our embeddedness in nature also has consequences for divisions within human relations: the meeting of basic needs, or "necessary" labor, is devalued by contrast with creative leisure pursuits that are in the domain of freedom. The links between this feature of Marx's thought and the feminist critique of the devaluation of caring, nurturing work are clear. Finally, humanist eco-Marxism is criticized for its continuing faith in the working class as the main agent of change, despite its incorporation into productivist priorities. Even Donald Lee's close approach to ecocentrism remains vulnerable to criticism on the grounds that the overcoming of alienation between humanity and nature takes place through a "humanization" of nature in which nature itself is rendered artefactual. However, even to get this close to an ecocentric perspective Lee has to abandon some of Marx's central concepts and distinctions.

Eckersley's conclusion is that there are elements of resource conservation and human welfare ecology in the work of Marx and Engels and their eco-Marxist followers. However, the full development of an ecocentric perspective would be incompatible with some defining features of Marxism. At the same time, she does acknowledge that "orthodox" eco-Marxism in particular continues to have relevance as an explanatory theory of ecological degradation.

However, the debate does not end here. In a subsequent chapter of her book, from which our second extract is drawn, Eckersley returns to a discussion of the work of ecologically informed socialists. Since "ecosocialists," as she defines the tradition, have dispensed with Marxist Promethean "productivism," and have also abandoned classical Marxist reliance on the working class as the sole agency of change, Eckersley sees them not as developers of the Marxist tradition but, rather, as "post-Marxist." Since the majority of contributors to this collection—and certainly most of those represented in Parts II and III—conform to Eckersley's definition of ecosocialism, but would probably still think of themselves as eco-Marxists, the "post-Marxists" label is clearly open to debate. However, there is probably little point in getting sidetracked on this matter of definition. Whatever we *call* this newly emergent result of the dialogue between Marxists and the proponents of political ecology, feminism, and other "new" social movements, Eckersley's description of its themes and values is a fair one. Her ecocentric critique is both interesting and powerful.

For Eckersley, two litmus tests of an ecocentric approach are attitudes regarding wilderness preservation and long-term human population reduction. Ecocentrics will tend to support these goals as enabling the flourishing of nonhuman nature while also promising benefits for humans in the longer term. Anthropocentrics will tend to oppose both, except insofar as humans can be shown to benefit in the short term. In other respects there is common ground between ecosocialists and ecocentric emancipatory politics, with the proviso that ecosocialists still tend to remain within an anthropocentric, though expansive and "enlightened," normative perspective. The challenge for ecosocialists is to show how their values of decentralized participatory democracy and civil liberties can be realized in ways that reject the bureaucratic centralism, state repression, and paternalism of earlier forms of "socialist" practice. The common ground between ecosocialists and ecocentrics is sufficient to justify a mutually beneficial effort at bridge building.

NOTE

1. Jonathan Porritt, *Seeing Green* (Oxford: Basil Blackwell, 1984), 44.

Chapter 10

ECOFEMINISM AND ECOSOCIALISM
Dilemmas of Essentialism and Materialism

MARY MELLOR

INTRODUCTION

The rethinking and recasting of historical materialism from an ecological perspective only takes account of one of the so-called "new social movements" of the late twentieth century. Another new, but in fact very old movement, feminism, must also be addressed. This certainly has been the intention of eco-Marxists such as the founders of the journal *Capitalism, Nature, Socialism (CNS).*[1] However, the integration of women as contributors to this debate in substantial numbers, or as theoretical subjects, has yet to be achieved. I hope that this chapter will contribute to what will necessarily be a long and complex debate. The core of my argument here is that it will prove impossible to construct an ecosocialist/ecofeminist revolutionary theory and practice unless we can finally break out of the laager of economic analysis to embrace women and nature not as objects of the economic system, but as subjects in their own right, an argument I have addressed more fully elsewhere.[2]

This endeavor raises the critical issue of the relationship between the seemingly ahistoric universals of biological sex and nature as "essential" features of human existence and the historical materialism of class analysis. Historical materialism asserts that the constraints on, and potential for, collective human development and creativity are socially constructed, and thereby capable of being socially resolved. The dilemma between essentialism and materialism is whether the socially materialist analysis of historical materialism can integrate the physically material reality of women and nature, that is, can we bring women in without their biology, or nature in without the constraint of its "natural" limits? These problems are inherent in the debate between ecofeminism and ecosocialism.

The debate in *CNS* was opened in response to the very cursory treatment of ecofeminism by Faber and O'Connor in their discussion of the environmental movement in the United States where they dismissed ecofeminism as fused with "neo-Romantic nature ideologies."[3] While agreeing that this criticism of some kinds of ecofeminism was valid, Lori-Ann Thrupp responded by claiming that ecosocialism was in danger of missing the "rich theoretical and historical analysis" of ecofeminism.[4] She pointed to the common exploitation that women and nature have received at the hands of patriarchy and the role that women have played in labor, antimilitary, and environmental struggles. Drawing on the work of Carolyn Merchant, Thrupp distinguished between the radical feminist grounding of human nature in human biology with its evocation of ancient rituals and goddess worship and socialist ecofeminism that sees both "nature and human nature as historically and socially constructed." While radical ecofeminist philosophy embraces intuition, an ethic of caring, and weblike human–nature relationships, socialist ecofeminism would seek to give both production and reproduction a central place in materialist analysis.[5]

Faber and O'Connor responded by apologizing for giving limited attention to ecofeminism but went on to affirm their criticism of "romantic" ideas such as intuition as against science and technology, the privileging of the human body over "mind," and "organic theories emphasizing emotional ties to the community ('caring')."[6] I do not think these ideas can be so easily dismissed; in particular, we cannot ignore emotional ties and caring if we are to theoretically integrate reproduction and production.

A central difficulty in discussing this relationship is that neither ecofeminism nor ecosocialism can be easily defined, reflecting the fact that the former draws on many feminisms and the latter on many socialisms. Ecofeminists range from New Age thinkers to socialists, and ecosocialists range from Marxists to anarchists.[7] However, it has become plain in the context of *CNS* that the debate is between a reading of ecofeminism that sees it as embodying variants of cultural or radical feminism and a neo-Marxian socialism:

> In the U.S., *radical* ecofeminism means more or less an "essentialist"
> view of women and men, not a "materialist" view of human nature as

socially and historically (as well as naturally) constructed, as in the socialist traditions that we associate ourselves with.[8]

The core of this argument is whether an analysis and programme for change based on sex/gender (reflecting relations of both reproduction and sexuality) stand independent of a programme for change based on production relations. Despite the best efforts of many theorists, far too numerous to list, a "marriage" between Marxism and radical feminism has not been achieved. Ariel Salleh is quite right to be concerned that a debate between ecofeminists and eco-Marxists might end up in a similar barren theoretical quagmire.[9] It is clear that Marxist socialism cannot "take account" of women-as-subjects and feminist theory generally without seriously threatening the male and productivist basis of Marxist theory.

At the same time, ecosocialists such as Faber and O'Connor who draw on a Marxist base quite rightly are concerned about any analysis, green or feminist, that threatens to divert attention from the massive and corrupting power of capital. The dilemma between essentialist readings of women and nature versus a materialist analysis of economic relations is that feminist and ecological concerns will undermine historical materialism by positing essentialist limits to human activity (ecological or biological) or by claiming an intuitive source of knowledge that draws on biological or ecological dynamics. From the perspective of historical materialism such a course would trap human societies in a reified naturalism whereby social relationships are presented as ordained by biology or by nature. The green arguments that there are "natural" limits to growth or that nature contains within it a "natural" balance, or claims by some feminists that they are "naturally" more peace loving and cooperative, risk presenting constructs of human society as constructs of nature.

I would argue that we will not overcome this dilemma by denying the *material* issues at the basis of the feminist and ecological critique. Rather, we should move to the central question of how we theorize the very real question of the finite nature of the planet and the biological differences between men and women. To maintain that there is a biological and ecological limit to human activity and our capacity for social reconstruction is not to revert to essentialism but to begin to theorize the conditions of our material existence.

GETTING DOWN TO ESSENTIALS

The extent to which the accusation of essentialism can be laid against ecofeminism depends on the way in which the relationship between women and nature is defined. Is it a relationship of affinity, of a unity of

spirit/biology between women and nature, or the sharing of a socially
constructed relationship of exploitation? Petra Kelly clearly expressed the
affinity of women with nature:

> Women are the "ombudsmen" [*sic*] of future generations . . . because
> only [a woman], I feel, can go back to her womb, her roots, her natural
> rhythms, her inner search for harmony and peace, while men, most of
> them anyway, are continually bound in their power struggle, the exploi-
> tation of nature, and military ego trips.[10]

As Cynthia Enloe has remarked, the overwhelming predominance of
men in international politics and the violent relations that ensue has meant
that there is scarcely a woman who "on a dark day" has not been attracted
by essentialist arguments about the inherently violent nature of men, if not
the inherently peaceful nature of women.[11] The same is probably true for
those women, like myself, whose political history lies in socialist movements. On my dark days I wonder if male-dominated structures are
irredeemably bureaucratic and sex/gender-blind. Given the very real
experience of women at the hands of men and the institutions they control,
it is inevitable that essentialist ideas tremble beneath the surface of most
feminist thought. This is particularly true of ecofeminism, one of whose
primary roots is in the women's peace movement.[12]

Do women have a privileged role in the "intricate web of life"
through the nurturing experience they share with the planet? Do they
have more peace-loving, intuitive, and caring characteristics? Green
writings that lean toward New Age thinking proclaim the existence of
feminine and masculine "principles."[13] The feminine principle associated with women exhibits the "soft/yin" qualities of cooperation, empa-
thy, holistic thinking, emotion, and intuition, while the masculine principle associated with men displays the hard/yang qualities such as
competitive assertiveness, rationalism, aggression, and materialism.
These principles are seen as timeless and universal but not essentially
limited to either sex. Ideally, each human being should be a balance of
yin and yang, of the masculine and feminine principles. Advocates of the
feminine principle seek a change of values so that we all become
"balanced" in a "cultural transformation" whereby men recover the
"feminine" side of themselves and women become more assertive. Cultural feminists have also proclaimed the superiority of women's culture,
rooted in her "nature"—although without any optimism about men's
ability to respond to women's "healing" powers.[14]

This form of analysis is clearly unacceptable to socialists. It returns to
exactly the kind of essentialist idealism that Marx and Engels opposed.
Historical materialism rejects such ahistorical universals and the claim that

changing (or recovering) values will change society. However, we are still left with a problem: while it is easy to reject idealist essentialism, the question of the relationship between the biological/ecological and the social remains. In the original "Prospectus" of *CNS* the "human–nature" puzzle is set out as one that has to be theoretically resolved.[15] This is a biological as well as an ecological puzzle. Murray Bookchin has proposed that we recognize a dialectical relationship between ecology and society in the concept of a "social ecology," but the idea of a "social biology" has a very different ring, largely because of the reactionary work of sociobiologists. However, the dangers of essentialism, and the reactionary nature of previous analysis, should not blind us to the need to incorporate the reality of biology and ecology into our theory and practice.

Not only socialists have been reluctant to open discussion about the "human–nature" puzzle; feminists also have only hesitatingly addressed the question of biology. As Ariel Salleh points out, the primary task of second-wave feminism was to overthrow the limitations that male-dominated conceptions of women's biology had placed upon them, and to deny that biology was destiny. At the same time, she points out that "the biological" cannot be thrust aside.[16] Motherhood, in particular, is being reassessed within feminist thought and forms a substantial basis of feminist analysis of the issues of ecology and peace.[17] Heather Jon Maroney argues that the failure to address the "presocial reality" of motherhood reflects a continuing desire on the part of men to maintain the separation of the public from the private.[18] Mary O'Brien sees it as a male-dominated urge to separate (social) life from (biological) necessity: "an ideological separation, a yearning and a dream of the sweet sunshine always outside the cave of the contradications of carnality."[19]

Salleh draws our attention to the danger of socialists perpetuating the Judeo-Christian, Baconian-Cartesian division between mind and body in ignoring the reality of women's biological experience: or "the *masculine* will to disconnect from and transcend our earthly condition: what Marx called 'necessity.' . . . the rationalist thrust to transcend bodily embeddedness in place and relationships."[20] Martin O'Connor points to the "inescapable materiality of pregnancy and childbirth."[21] O'Brien argues that the separation of production and reproduction that runs right through the work of Marx and the Marxists translates what is specifically a male experience of that separation into a false universal truth. I would argue that the claim that biology/ecology can be subsumed within the human determination of the social is unrealistic and male-oriented in its prioritizing of the economic over all other aspects of human and nonhuman existence. In effect, Marxian socialism is presenting us with a *normative* theory that represents the experience and ideology of men. In this it *shares* rather than opposes the perspective of capital.

WHAT IS MATERIAL ABOUT
HISTORICAL MATERIALISM?

If we are to synthesize ecosocialism and ecofeminism we must be clear about the relationship between historical materialism and the material relationship between men and women. More contentiously, perhaps, we need to theorize the way in which biological differences are reflected in male–female relations. This brings us to the crux of the question of the relationship between what we might call physical materialism and social materialism. This is caught in Marx's epithet that "[men] make history, albeit not under conditions of their own choosing." How far are human biological differences (and ecological limits) not "conditions of our own choosing"? The foundation of Marxist historical materialism is the primacy of economic relations. Under capitalism, these are primarily relations among men and, despite valiant efforts from Engels onward, the position of women has never been adequately theorized within historical materialism.[22]

In Marxist theory, economic relations are based on the primary need of human beings to produce the means of their survival. In particular eras this takes the form of a particular mode of production of which the most world-dominant has been capitalism. However, what does it mean to say that economic relations are "determinant in the last instance," to use Althusser's phrase? Is this not in itself an essentialist statement? To Marxists this question is sacrilege, but from a feminist point of view a challenge must be made. In *The German Ideology,* in Marx's discussion of the social relations of production, he talked first of "the production of life, both of one's own labor and of fresh life by procreation." Why should the means of survival (a biological imperative) be allowed into historical materialism but not the means of reproducing life itself? Further, if the means of survival produced definite social relations and particular forms of consciousness, why not the means of procreation?

I would argue that there are three interlinked material bases to human society: the forces/relations of production, the forces/relations of reproduction, and the relations between human society and nature. In the case of the first two the distinction between them is in itself socially constructed. What is incorporated in the sphere of "production" does not just represent the interests of capital, it represents the interests of men. By separating production from both reproduction and from nature, patriarchal capitalism has created a sphere of "false" freedom that ignores biological and ecological parameters. It is a sphere that can exploit nature without paying attention to what O'Connor has called the "second contradiction of capital," the conditions of production itself.[23] However, unlike O'Connor, I do not think this is a contradiction just for capital: it is a contradiction for men as well.

The integration of ecofeminism with ecosocialism cannot be achieved by trying to add women onto a male-dominated productivist socialism; that has been tried many times and failed lamentably. Marxian socialism must be reconstructed to take account of the reality of women's lives and the way in which the male/capitalist sphere of production is materially dependent upon women and nature. The reality of women's lives is not only sexual and reproductive, but productive, particularly in the struggle for livelihood by the women of the South. As the women of the Development Alternatives with Women for a New Era (DAWN) point out, "In food production and processing, in responsibility for fuel, water, health care, child-rearing, sanitation and the entire range of so-called basic needs, women's labor is dominant."[24] The invisibility of women's work and its existence outside of formal economic relations still has to be placed on the (male) political agenda.[25] However, capitalism has found women's caring and domestic work and builds upon preexisting patriarchal structures to produce appalling hardship where women are exploited as workers and as women, to the point of "superexploitation" where even their basic subsistence is denied.[26]

In women's lives, particularly in subsistence economies, it is impossible to separate productive and reproductive work. It is this work of producing the means of life and of survival, ecofeminists argue, that establishes the close relations between women and the planet. Vandana Shiva claims that in their subsistence work women have been guardians of the ecological sustainability of the planet, and that woman-based sustainability is now threatened by the cash-cropping and genetic engineering of patriarchal capitalism.[27] It is in women's lives that the relationship between the social/biological is constructed and it is this underlaboring work that women do that is not incorporated into the "material" world of men as represented in the theoretical framework of historical materialism. By that token, overthrowing capital will not resolve the "second contradiction" of the conditions of production for either women or nature. In fact, the theory of historical materialism based on the primacy of (formal) economic relationships as defined by capitalism has produced the catastrophe of the command economies which, by mimicking the productive power of capital, continued to deny the reality of much of women's lives and produced ecological devastation.

WOMEN'S LIVES: NURTURING THE WORLD

Feminists are increasingly focusing on the sexual divisions that surround mothering and nurturing and the role of women in sustaining human physical and emotional existence as a means of understanding our relationship with the natural world and as the basis of a reconstructed socialism:

> Because of the way it mediates between the biology of procreation and historical institutionalization, motherhood provides a prime site for exploring and constructing boundaries between nature and culture. Historically, the division in Western thought has been dichotomous and drawn in such a way as to exclude women from the social and the historical.[28]

> The world of nurturance and close human relationships is the sphere where the basic human needs are anchored and where models for *humane* alternatives can be found. This world, which has been carried forward mainly by women, is an existing alternative culture, a source of ideas and values for shaping an alternative path of development for nations and all humanity.[29]

This is not to return to an essentialist idealization of "women as mothers." Not all women are mothers or want to be mothers. Not all mothers enjoy that role. Moreover, "mothering" is a role that can be carried out by men. The potentially positive values of mothering attach to the performance of the task, not to the biology of the performer.[30] The fact that women biologically can and do give birth does not imply any particular adoption of feminine/feminist values. Women who have borne children do go to war and do embrace militaristic nationalism, racism, and fascism. However, with Ann Ferguson and Sara Ruddick, I would argue that mothering/nurturing is a biological/social task that has to be carried out if human society is to survive and that in a patriarchal society that task is carried out, willingly or unwillingly, by women.

Women's work as mothers and nurturers, or their identification with that role, has justified their marginalization within economic relationships to the extent that their work, quite literally, does not count. Throughout history this work has led to the association of women with nature rather than with culture,[31] constraining them to the private sphere of unfreedom, while men colonize the public "sphere of freedom" untrammeled by domestic necessities. Although sexual and reproductive relationships have been constructed differently across history and across cultures, in their domestic, nurturing, and subsistence work women have been held almost universally responsible for meeting the immediate emotional and material needs of their families and for sustaining relationships within the community.

The main characteristic of women's work, as Charlotte Perkins Gilman pointed out, is its "immediate altruism." Women's work is altruistic in the sense that it is carried out for only incidental personal gain (the pleasure of close personal relationships), and it is immediate in the sense that it cannot be "put off" or slotted into a work schedule. Immediate emotional and physical care cannot be "logically" ordered or "rationally programmed." The needs to which women respond are demands that cannot be ignored; if they are ignored, the social fabric of society begins to disintegrate: "The main distinction of human virtue is . . . altruism [and] 'otherness' . . . to love and serve one another, to feel for and with one another. . . . The very existence of humanity is commensurate with [the] development [of them]."[32]

While the work that women do is both socially and biologically essential, women's responsibility for meeting the immediate needs of both family and community is not a biological given, nor is it freely chosen: it is an "altruism" imposed on women by men. I would argue that this is the most fundamental division in society, that women are primarily (but not all and not always) responsible for meeting the immediate needs of others, while men (again, not all and not always) are not. In industrial or market economies, a great deal of women's subsistence work and some caring work is incorporated within the economic system, but, if that system fails, women once more begin the long struggle of "holding the family together" and maintaining the relationships that create "humanity." In doing this work, women produce a substantial proportion of the material basis of men's lives, but perhaps more importantly give men that most precious asset: time.[33] Ignoring women's role in the production of time produces a public world that does not take account of the reality of women's lives. As feminists have pointed out, the division between the public and the private world is a false one that reflects male interests and male experience. It is a world that can compartmentalize human existence into categories of space and time, where decisions can be made that do not take account of the complexities of human existence. It is a world where military, scientific, or economic "logic" can be pursued regardless of its impact on human relationships or even human existence.

The combined and uneven development of men's and women's lives was exacerbated by industrial capitalism, but it was not created by it. Patriarchal society has a long history whose origins are still obscure.[34] Failure to recognize women's work means that male-dominated capitalism is constructed on a false premise, that of the independently functioning individual. This reflects not only bourgeois individualism but patriarchal individualism, reflected in economic terms (Marxist and non-Marxist) as "free" labor. Only a small minority of men, and an even smaller minority of women, actually achieve sufficient power to function independently, but that does not prevent the public world from being constructed on that basis. Women who enter the public world have to operate according to the principles of male/bourgeois individualism, that is, they must deny any domestic responsibilities or pass them on to someone else (usually another woman). Men who undertake caring responsibilities must also be available to work long hours and be ready to uproot themselves and their families according to political/economic or military demands.

Insofar as Marxist theory shares its definition of economic relations with patriarchy and capitalism, it will be unable to break through the theoretical barriers of its own construction. Marxist socialism rejects the capitalist economic relations that exploit workers, peoples of the South, women, and the planet, but it does not recognize that it shares with these relations the artificial boundaries of male-dominated productivism, so that women's lives become theoretically a leftover category, the "sphere of reproduction." The differences in men's and women's approaches to life are not defined by some biological "essentialism," nor do they reflect universal male or female "principles"; rather, they reflect the very real differences in life experience of men and women, male-experience-reality (ME-reality) as against women's-experience-reality (WE-reality). By the same token, a feminist analysis that focuses upon women's experience is no more essentialist than the Marxist emphasis on the predominantly male experience of the public world of production. We are not talking about a distinction between essentialist versus materialist theories, but rather a particular dynamic between the essential needs of human existence and the material construction that is put upon them. The essentialism–materialism dichotomy is, in fact, a contradiction. It *appears* as a dichotomy if viewed through male-defined reality; but when viewed from the perspective of women's lives, the dichotomy can clearly be seen as a contradiction. The male construction of a social world presupposes its material base in women's time and work. When women try to articulate a perspective that reflects their social condition, they are accused of essentialism, or at least of detracting from the "primary" economic struggle with capital.

In the debate between essentialism and materialism we must be sure whose reality we are representing. A socialism that starts from ME-reality prioritizes public relationships, in particular, economic relationships. A socialism constructed on the basis of ME-reality will be *essentially* limited. It will establish itself on a limited material base, an economic system defined by both patriarchy and capitalism.

A FEMINIST REALITY

Like Ariel Salleh, I was concerned to see Faber and O'Connor from an eco-Marxist perspective dismiss "organic theories emphasizing emotional ties to the community ('caring')" as "romantic" radical feminism.[35] Women's lives are caught in a network of interconnected relationships not as an essentialist ideal, but as a material reality. Their thinking has to be feeling, intuitive, and multifaceted because that is the structure of their lives. Dubbing ecofeminism "essentialist" means that it can be easily marginalized and rejected by socialists without a real engagement with feminist theory and practice. There is no question but that both women and the planet have suffered abominably at the hands of men, and that men have a case to answer. The problem is to explain the dominance of men in destructive practices without collapsing into an equally one-sided feminism.

The answer lies in the material limitations of the need to secure human reproduction and survival and the parameters it places upon human (in practice, women's) activity. Failure to acknowledge these limitations masks the material benefit that men gain, as men, from women's work.[36] A feminist ecosocialism must not dodge the fact that there is a direct connection between the biological differences between men and women and the social construction that is put upon them: between the forces and the relations of reproduction. Biological differences of sex do not determine human behavior; they are the forces of reproduction that have to be accommodated in relations of reproduction. To put it more simply, in the absence of the reproductive technology that Firestone envisaged (and many feminists fear men are trying to create), women will continue to bear children and be primarily responsible for them, at least in the early years.[37]

This is not to say that biology determines the power relationships between men and women, but it does constrain them. To resolve those power relations will not mean that biological differences between the sexes will wither or be "willed away." I would agree with Pat and Hugh Armstrong that the distinction between sex and gender, while helpful in many ways, ultimately can be misleading. Women will continue to be subordinated under socialism "unless socialist women and men are prepared to take their biological differences into account."[38] As Ynestra King put it, there is a "masculinist mentality which would deny us our right to our own bodies and our own sexuality, and which depends on multiple systems of dominance and state power to have its way."[39] Recent concern about reproductive engineering reminds us that scientific developments are not just about creating profits by genetic engineering, they are also about men controlling women through their biology.[40] It is also not essentialist to point to the implications of the fact that men, as a rule, do not mother. Dorothy Dinnerstein and Nancy Chodorow have both warned convincingly about the dangers of mothering being assigned exclusively to women.[41] From a psychoanalytic perspective, they argue that girls respond to female mothering by becoming dependent on emotional relationships and continue to identify with the mothering role, while boys have to distance themselves from emotional relationships in order to follow the distant father into the public world where, in their emotionally truncated state, they can wreak havoc. Carol Gilligan has argued that the differences in male and female development have created two different moral universes where "male and female voices typically speak of the importance of different truths, the former of the role of separation as it defines and empowers the self, the latter of the on-going attachment that creates and sustains the human community."[42] Elizabeth Spelman has quite rightly pointed out that Freudian theory, from which Dinnerstein and Chodorow draw their inspiration, is historically and culturally specific and cannot be assumed to apply universally to family and child-rearing, but this does not detract from its applicability to male and female development in Western cultures.[43]

There is no biological reason why men should not nurture children after birth, and many do. However, the reason most men do not "mother" is not purely social; it is far too universal a phenomenon for that to be the case. I would suggest that a possible explanation is social inertia resting on a biological phenomenon. If women give birth, and in most cultures suckle, what is there to encourage men to take over nurturing at some later stage? I would argue that only a socialism that recognizes the *fact* of biological difference, and constructs a political response to it, will be able consciously to take control of human history and found a society based upon WE-reality. Feminists have argued that their biological/social position in society has given women a specific standpoint that enables them to produce an alternative view of the world, one that "can overcome the old, oppressive dichotomy between the natural and the social," that can "represent a more complete materialism, a truer knowledge."[44] As Nancy Hartsock has pointed out, "The lived realities of women's lives are profoundly different from those of men." It is from the perspective of women that we can transcend the false boundaries between the natural and the social and accept that "as embodied humans we are . . . inextricably both natural and social."[45]

A feminist standpoint would lead us away from the ME-reality of the male-oriented public world and toward a WE-reality society that would find space and time not only for our physical needs, but also for emotional sustenance and the development of nonmaterial aspects of human development. Only a socialism that not only incorporates, but prioritizes, women's experience will be able to create the kind of society that will relieve women of the imposed altruism of their nurturing and caring work. By reintegrating the divided worlds of men and women, such a society will pull men back to the pace of biological time in which women live. It will create a society in which political and economic decisions will be local and accountable. It will equalize the resource of time, so that we can begin to slow down the pace of human development to sustainable levels. This is not to deny the present overwhelming dominance of capitalist economic relations. Theory or practice based on WE-reality will not of itself confront multinational companies, end the nuclear arms race, or prevent the exploitation of workers. An ecofeminism that does not embrace socialism would be as theoretically and politically limited as an ecosocialism that does not embrace feminism.

TOWARD A FEMINIST GREEN SOCIALISM

A feminist green socialism would not assume that human society is so exclusively determined by its mode of production that all other so-cial/biological/ecological structures will "fall into line" if that mode of production is changed. It would accept that while there is no predeter-mined destiny in biology or ecology, we need to come to terms politically with their reality. It is true that to see aspects of human existence as a "given" prevents us from realizing that, through collective struggles, we can reconstruct our *social* world on egalitarian principles. And, quite rightly, to defend this principle socialists have waged a fierce battle against all forms of naturalism and essentialism, against all claims that certain aspects of human existence are beyond social control, and can only be "discovered," not constructed. However, in defense of the social we must not overly socialize the natural. To do so obscures the ecological framework of our existence and leads to a mystification of the material conditions of women's lives. The dominance of economic relationships in Marx and Engels's thought was, to say the least, unfortunate given the importance of feminist analysis in early French and British socialism.[46] Marx and Engels were well aware of the embeddedness of humanity in nature and of nature in humanity, as well as of women's inequality, but neither were incorporated into Marx's critique of capital. Engels at-tempted to construct a dialectics of nature and explain the origins of the patriarchal family, but neither subject has been substantively taken up by the legions of predominantly male, middle-class, and white Marxist theorists, nor have they formed part of the structure of socialist theory.

Faber and O'Connor criticize "some self-described ecofeminist prac-tice" for opposing "some self-described ecosocialist practice as irredeem-ably 'patriarchal.' "[47] I see this as something male ecosocialists must refute, rather than as something ecofeminists should justify. Socialism, particu-larly in its Marxist form, has been "irredeemably" patriarchal for most of its existence. A feminist ecosocialism will need to expand historical mate-rialism to explain relationships other than economic ones. Moreover, it will need to question the reification of the economic. It will have to recognize that economic relationships are not just defined by capitalism, but are also defined by patriarchy. Look, for example, at production decisions in Eastern Europe, guns and heavy machinery, but no adequate contraception, anesthetics for abortions, and limited domestic equipment and cleaning materials. A feminist green socialism would also recognize that men have very real interests in controlling women's sexuality and the domestic relations of production and reproduction, with their reflection in occupational segregation, inequalities, and sexual harassment.

By reifying economic relations, Marxian socialist theory has been limited by the boundaries of the capitalist mode of production. It has itself suffered from essentialism. A feminist ecosocialism would insist that the socially constructed "economy" should not be seen as the sole determining "material reality." Other realities are equally important, especially those of women and nature, and cannot be accommodated without reconstructing the whole socialist project. Socialists must realize that the boundaries of women's lives are not defined by economic relations: women cross and recross the so-called public–private world. Most important for women is the fact that the boundaries of their lives are policed by male violence, which also crosses the public–private divide. This is something that male-dominated socialism has not even begun to take on board.

At present, Marxist theory reflects the historically specific structure of Western patriarchal capitalism which prioritizes the interests of some men (and a few women) against the remaining men, exploited and oppressed by race and class, and the vast majority of women. For women, overthrowing those structures means accepting the biological materialism of sex differences and building a society that prevents social inertia starting the whole destructive process all over again, with men colonizing the "public" world and confining women to "inferior" work. A feminist ecosocialism would not deny the exploitative reality of capitalist economic relations or class politics, but would see it as one aspect of a much wider set of material relations.[48] A socialism conceived in the industrial center of a colonial power, based on a dominant race and sex, is not equipped to transport us to a feminist or a green future. If we have a socialism that only recognizes and embraces the male-dominated world of trade, war, and politics, a socialist revolution will do nothing to reconstruct those priorities. Men will continue to "take women's time," relying on women to be available for the twenty-four-hour responsibility of creating the framework of physical, emotional, and social relationships that make human society possible. They will be able to continue designing weapons of war, economic structures, and political systems that discount the needs of the majority of the world's peoples. A socialism that does not challenge the economic and sexual domination of women by men will never achieve an egalitarian society, nor one that is ecologically sustainable.

ACKNOWLEDGMENTS

The author wishes to thank Nancy Folbre for her superbly helpful editorial support, and Dr. Barbara Holland-Cunz, Jack Kloppenburg, Jr., Ropo Sekoni, and Arun Agrawal for their many useful comments.

NOTES

1. "Discussion" between Lori-Ann Thrupp, Daniel Faber, and James O'Connor, in *CNS*, Issue 3, November, 1989; "Discussion" between Ariel Salleh, Martin O'Connor, James O'Connor, and Daniel Faber, *CNS*, 2(1), Issue 6, February, 1991.

2. Mary Mellor, *Breaking the Boundaries: Towards a Feminist, Green Socialism* (London: Virago, 1992); see also Brinda Rao, *Dominant Constructions of Women and Nature in Social Science Literature* (Santa Cruz, Calif.: CES/CNS Pamphlet 2, 1991); Mary Mellor, *Feminism and Ecology* (Polity Press), forthcoming.

3. Daniel Faber and James O'Connor, "The Struggle for Nature: Environmental Crisis and the Crisis of Environmentalism in the United States," *CNS*, Issue 2, 1989, 32.

4. Thrupp, "Discussion," November, 1989, 170.

5. Ibid., 172–173.

6. Faber and O'Connor, "Discussion," February, 1991, 177.

7. See, for example, the myriad themes and visions woven into two recent ecofeminist anthologies: Irene Diamond and Gloria Feman Orenstein, eds., *Reweaving the World* (San Francisco: Sierra Club Books, 1990); and Judith Plant, ed., *Healing the Wounds: The Promise of Ecofeminism* (London: Green Print, 1989). For a review of socialist perspectives, see Martin Ryle, *Ecology and Socialism* (London: Radius, 1988).

8. Faber and O'Connor, "Discussion," February, 1991, 138; italic in original.

9. Salleh, "Discussion," February, 1991, 134.

10. Petra Kelly, *Fighting for Hope* (London: Chatto and Windus, 1984), 104.

11. Cynthia Enloe, *Bananas, Beaches, and Bases* (London: Pandora 1989), 5.

12. Leonie Caldecott and Stephanie Leland, eds., *Reclaim the Earth* (London: Women's Press, 1983).

13. Fritjof Capra, *The Turning Point* (London: Flamingo, 1982); Dorothy and Walter Schwarz, *Breaking Through* (Bideford, U.K.: Green Books, 1987). Ecofeminists have also expounded the idea of a feminist principle; see, for example, Stephanie Leland, "Feminism and Ecology: Theoretical Considerations," in Caldecott and Leland, eds., *Reclaim the Earth*.

14. Mary Daly, *Gyn/Ecology: The Metaethics of Radical Feminism* (London: Women's Press, 1989); Andree Collard, with Joyce Contrucci, *Rape of the Wild* (London: Women's Press, 1988); Susan Griffin, *Women and Nature* (London: Women's Press, 1984).

15. *CNS*, Issue 1, 1988, 3.

16. Salleh, "Discussion," February, 1991, 130.

17. Sara Ruddick, *Maternal Thinking* (London: Women's Press, 1989).

18. Heather Jon Maroney, "Embracing Motherhood: New Feminist Theory," in *Politics of Diversity*, eds. Roberta Hamilton and Michele Barret, (London: Verso, 1986).

19. Mary O'Brien, *The Politics of Reproduction* (London: Routledge and Kegan Paul, 1981), 141.

20. Salleh, "Discussion," February, 1991, 134; italic in the original.

21. Martin O'Connor, "Discussion," February, 1991, 136.

22. Zillah Eisenstein, ed., *Capitalist Patriarchy and the Case for Socialist Feminism* (New York: Monthly Review Press, 1979); Heidi Hartmann, "The Unhappy Marriage of Marxism and Feminism: Towards a More Progressive Union," in *Women and Revolution* ed. Lydia Sargent (London: Pluto Press, 1981); Michelle Barrett, *Women's Oppression Today* (London: Verso, 1980); Annette Kuhn and AnnMarie Wolpe, *Feminism and Materialism* (London: Routledge and Kegan Paul, 1978); Lise Vogel, *Marxism and the Oppression of Women* (London: Pluto Press, 1983).

23. James O'Connor, "The Second Contradiction of Capitalism: Causes and Consequences," in *Conference Papers* (Santa Cruz, Calif.: *CES/CNS* Pamphlet 1, 1991).

24. Gita Sen and Caren Grown, *Development, Crises, and Alternative Visions* (New York: Monthly Review, 1987), 23–24. See also Vandana Shiva, *Staying Alive* (London: Zed Press, 1989); and Brinda Rao, "Struggling for Production Conditions and Producing Conditions of Emancipation: Women and Water in Rural Maharashtra," *CNS*, Issue 2, Summer, 1989.

25. Marilyn Waring, *If Women Counted* (London: Macmillan, 1989).

26. Maria Mies, *Patriarchy and Accumulation on a World Scale* (London: Zed Press, 1986).

27. Shiva, *Staying Alive*, 96ff.

28. Maroney, "Embracing Motherhood," 398.

29. Hilkka Pietilä, "Alternative Development with Women in the North" (Paper given to Third International Interdisciplinary Congress of Women, Dublin, July 6–10); also published in Johan Galtung and Mars Friberg, eds., *Alternativen Akademilitteratur* (Stockholm, 1986).

30. Ann Ferguson, *Blood at the Roots* (London: Pandora, 1989); O'Brien, *Politics of Reproduction*, 22; Ruddick, *Maternal Thinking*, 52.

31. Sherry Ortner, "Is Female to Male as Nature Is to Culture?," in *Woman, Culture and Society*, ed. Michele Z. Rosaldo and Louise Lamphere (Stanford, Calif.: Stanford University Press, 1974).

32. Charlotte Perkins Gilman, *Women and Economics* (London: G. P. Putnam's Sons, 1915), 523.

33. Frieda Johles Forman, ed., *Taking Our Time* (Oxford: Pergamon, 1989); Mellor, *Breaking the Boundaries*, 249ff.

34. Mies, *Patriarchy*, 44ff.

35. Faber and O'Connor, "Discussion," February, 1991, 138.

36. Christine Delphy, *Close to Home* (London: Hutchinson, 1984).

37. Shulamith Firestone, *The Dialetic of Sex* (London: Women's Press, 1979).

38. Pat Armstrong and Hugh Armstrong, "Beyond Sex-less Class and Class-less Sex: Towards Feminist Marxism," in *Politics of Diversity,* ed. Roberta Hamilton and Michele Barret (London: Verso, 1986), 252.

39. Ynestra King, "The Ecofeminist Imperative," in Caldecott and Leland, *Reclaim the Earth,* 10.

40. Gena Corea, *The Mother Machine: Reproductive Technologies from Artificial Insemination to Artificial Wombs* (New York: Harper and Row, 1985).

41. Dorothy Dinnerstein, *The Mermaid and the Minotaur* (New York: Harper and Row, 1976); Nancy Chodorow, *The Reproduction of Mothering* (Berkeley and Los Angeles: University of California Press, 1976).

42. Carol Gilligan, *In a Different Voice: Psychological Theory and Women's Development* (Cambridge, Mass.: Harvard University Press, 1982), 156.

43. Elizabeth V. Spelman, *Inessential Woman* (London: Women's Press, 1988).

44. Hilary Rose, "Beyond Masculinist Realities: A Feminist Epistemology for the Sciences," in *Feminist Approaches to Science,* ed. Ruth Bleier (London: Pergamon Press, 1986), 72.

45. Nancy Hartsock, "The Feminist Standpoint: Developing the Ground for a Specifically Feminist Historical Materialism," in *Feminism and Methodology,* ed. Sandra Harding (Milton Keynes, U.K.: Open University Press, 1987).

46. B. Taylor, *Eve and the New Jerusalem* (London: Virago, 1983).

47. Faber and O'Connor, 1989, "Struggle for Nature," 177.

48. See, for example, Ariel Salleh, "Nature, Woman, Labor, Capital: Living the Deepest Contradiction," in *Is Capitalism Sustainable?,* ed. Martin O'Connor (New York: Guilford Press, 1994).

Feminism, Ecosocialism, and the Conceptualization of Nature

KATE SOPER

Mary Mellor is quite right when she notes that a Marxist/socialist framework cannot "embrace women and nature" without itself being transformed in the process. The ecological and feminist agendas are not to be tacked on to an unchanged core of historical materialism, but can only be integrated into a Marxist perspective that has exposed and rejected its productivist, gender-blind, and class-reductionist tendencies. I am in agreement with her overall diagnosis, and welcome her discussion both for the forcefulness with which it states the case for a more genuinely integrative approach and for the light it sheds on the problems that stand in the way of this.

The points that I would like to take up do not concern the general nature and direction of her analysis, but instead relate to areas of her argument where I am uncertain of her position or feel it to be ambivalent.

In the first place, I sensed a prevarication around the "woman–nature" connection. This seems to underpin a good part of her argument, even though she is (rightly in my opinion) critical of various currents of ecofeminist romanticism. On the one hand, she warns us against any "essentialist" supposition of a spiritual affinity between women and nature; on the other hand, she offers a number of approving comments on those writers (e.g., Kelly, Gilligan, etc.) whose own arguments might seem to support this idea; and Mellor herself tends to elide "women" with "nature," with "biology," and with "mothering" in a way that I find both ideologically difficult and conceptually confusing. I find this alignment ideologically difficult not only because of the extent to which it has served to legitimate the confinement of women to maternal and nurturing roles, but also because insofar as "nature" has been coded "other" to a distinctively "human" culture (which has perennially both exploited and exalted it), the association of femininity with naturality tends to perpetuate rather than challenge an implicitly masculinist conception of "humanity."

From a conceptual point of view, I think it is best to avoid using the terms "nature" and "biology" as if these directed us to a female side of politics (concerned essentially with "reproduction") when they are better seen as referring us to properties and processes that are common to all animate and inanimate entities. Men, in this sense, are just as clearly situated within "nature" and governed by biology as women are, and both men and women are clearly involved in reproduction. This is not to deny the differential roles of the sexes in the process, or the sexual division of labor that has been culturally erected around it. But I think that to approach reproduction as if it were inherently a more "natural" and "female" domain obscures the politically important discriminations that need to be made with respect to *all* human practice (both "productive" and "reproductive") between those aspects that are biologically/naturally constrained (and thus far "essentially" given), and those that are socioeconomically conditioned, and whose specific "material" constraints are in principle removable.

These points connect with the uncertainty I felt about how far Mellor is calling on Marxists/socialists to change their values and priorities in the light of the centrality and importance of reproduction and female responsibilities with regard to reproduction, or how far she is calling for a transcendence of the cultural arrangements that have made nurturing and parenting so exclusively a female affair. At times, Mellor seems to be suggesting that a properly "materialist" analysis would recognize the role played by biological sex difference in making women the primary caretakers of the young, at least in their early years, and that a socialist politics must take account of this reality and integrate it into its analysis and practical programs of action; but at other times—in her suggestions that "mothering" and the caring ethic need not be exclusive to women, and her support for coparenting—she seems to imply that a properly "materialist" approach must target the exploitation of women that has been sustained through their cultural construction into the nurturing role, and its political program must be directed to dismantling this conventional division of labor. How far, in short, is Mellor seeking to relieve women of their "mothering" duties, and how far is she demanding a revaluation of them? Her discussion of the "altruism" imposed on women by men tends to suggest that she regards the former as most important, but it is the latter demand that seems to be implied by her points about the "not purely social" reasons why men don't "mother." So it is that when she concludes that "only a socialism that not only incorporates, but prioritizes, women's experience will be able to create the kind of society that will relieve women of the imposed altruism of their nurturing and caring work," it is difficult not to feel that, although this is presented as some sort of solution, it is actually more equivocal than she recognizes.

Finally, I am a bit unhappy about the opposition between "essentialism" and "materialism" that Mellor draws on in the early part of her discussion (though I do note the points she makes about the "contradictory" nature of the dichotomy). It seems that a materialist account must necessarily be "essentialist" at a certain level. It must recognize a distinction between "deep" and "surface" concepts of "nature," a distinction between the constant causal powers and processes comprising the deep structure of physicality, and the empirically observable "nature" (the environment, the body) that is continually transformed as a consequence of the operation of those causal powers. The contrast here is between "nature" as a presupposed and permanent ground of all human activity and environmental change, and the "nature" through which we refer to the historically changing and culturally transformed environment, or to the "lived experience" of physiology.

Rather than opposing "essentialist" and "historical materialist" approaches, I think we should recognize that the coherence of the socialist ecological critique of the cultural practices adopted toward "nature" depends on observing an essentialist concept of "nature." But it is also important to recognize that although nature in this essentialist conception does set limits on all human activity, however Promethean in ambition, these limits are extremely elastic, and are, in fact, "respected" in any and every form of ecological activity, even that most deleterious to planetary survival or human well-being. To be essentialist about nature, in this sense, is to acknowledge how *under*determining nature is of either ecological practices or gender arrangements.

I make these points not because I imagine Mellor would necessarily disagree with them, but because of their critical bearing on the rather indiscriminate, and potentially confusing, use of the term "nature" in much ecological and feminist writing, including some of Mellor's own formulations. For example, the implied contrast between the "natural" and the socially constructed is misleading to the extent that much ecological talk about "nature" is referring us to a cultural product. The "nature" that we endorse or disdain, from which we are "distanced" or have "affinity," is not nature as physical law and process. The use of the terms "nature" and "natural" to legitimize or prescribe a certain political agenda needs to be distinguished from its use as an explanatory-descriptive term. Those who make injunctions to "embrace" ("conserve," "respect," etc.) "nature" do, I think, need to be aware of these complexities in the use of the term rather than rely on its seemingly "obvious" meaning.

Chapter 11

SOCIALISM
AND ECOCENTRISM
Toward a New Synthesis

Robyn Eckersley

Extract 1, from Eckersley's Chapter 4:
The Ecocentric Challenge to Marxism

The recent efforts to develop a Marxist solution to the environmental crisis may be divided into two streams in accordance with the convenient distinction between "humanist" and "orthodox" Marxism (which loosely maps onto the work of the "young" and of the "mature" Marx, respectively). The humanist eco-Marxists have sought to develop a more ecologically sensitive Marxist response to the environmental crisis that seeks to harmonize relations between the human and the nonhuman realms. Orthodox eco-Marxists, on the other hand, make no apologies for being anthropocentric and are critical of humanist eco-Marxists for being idealist, voluntarist, and decidedly "un-Marxist." From an ecocentric perspective, however, it can be shown that *both* streams of eco-Marxism uncritically accept Marx's view of history and his particular notion of humanity as *homo faber* and thereby perpetuate an instrumentalist and anthropocentric orientation toward the nonhuman world.

This chapter consists of slightly revised extracts from Robyn Eckersley's influential book *Environmentalism and Political Theory*. Some of the wider context of these texts is presented in the Introduction to Part IV of this book. (T. B.)

ORTHODOX ECO-MARXISM

Orthodox eco-Marxists have strayed very little from the basic position of the "mature" Marx.[1] That is, environmental problems, like social problems, are traced directly to the exploitative dynamics of capitalism. The solution to these problems is seen to require the revolutionary transformation of the *relations* of production combined with the development of a better theoretical understanding of nature and further advances in technology so that a complete social mastery of nature can be attained for the benefit of all, rather than just for the privileged capitalist class. The orthodox eco-Marxist interpretation retains this Marxist view of history as a progressive dialectical struggle from the primitive to the advanced, resulting in the increasing domestication of the nonhuman world through the activity of labor and its extension, technology. In view of this, as Tolman explains,

> it should be clear why Marxists should continue to support the development of science and technology, and why they should assert the ultimate unity of science, technology, the mastery of nature, and humanism. Taking human history as a dialectical whole, these can all be seen as essential components of human nature itself. If understood in this dialectical sense, the Marxist gladly accepts the charge of "homocentrism."[2]

To orthodox eco-Marxists, the setting aside of areas of wilderness for the protection of endangered species and the preservation of biotic diversity in general can be justified only if it can be shown to be of some *instrumental* value to humans, by providing, say, a place of recreation or a store of potential raw materials for humanity's future productive labor. According to orthodox eco-Marxists, it simply makes no sense to say that the nonhuman world *ought* to be valued and protected for its own sake. For example, Howard Parsons, in his exhaustive review (and endorsement) of Marx and Engels's position on nature and technology, has trouble in grasping what the case against anthropocentrism is all about: "It is hard to know," he confesses, "what could be meant by nature 'in itself,' either in a Kantian sense or in the sense of a discrete reality entirely independent of our cognition and action."[3] On the basis of this epistemological point—that nature cannot exist independently from our values and actions—Parsons concludes that humans cannot *value* the nonhuman world for its own sake.

Yet Parsons's answer to the critique of anthropocentrism misses entirely the *normative* point of the ecocentric critique. First, he commits what Fox refers to as "the anthropocentric fallacy" in that he conflates the trivial and tautological sense of the term "anthropocentric" (i.e., that all our views are, necessarily, human views) with the substantive and informative sense of the term anthropocentric (the unwarranted, differential treatment of other beings on the basis that they do not belong to our *own* species). Second, and in any event, the ecocentric argument is not that we should value the nonhuman world because it exists independently from human values and actions. Ecocentric theorists would be the first to agree that there are no absolute divides in nature, that we are connected, in varying ways, with the nonhuman world and vice versa. However, despite these interconnections, the model of internal relations that informs eco-centrism also recognizes the *relative autonomy* of all entities. On the basis of this recognition, ecocentric theorists are concerned to cultivate a prima facie orientation of nonfavoritism that allows both human and nonhuman entities to unfold in their own ways. And it is precisely because we are part of an interconnected, larger whole that ecocentric theorists argue that we should exercise our own freedom with care and compassion.

However, it is clear from Parsons's discussion that even if he properly grasped the ecocentric argument he would still reject its normative and practical claims. For example, Parsons rejects as "unrealistic" the radical ecological argument that we should simplify human needs, reduce human population and consumption, respect nature, and lead a more agrarian life-style. (All but the last of these points more or less reflect the kinds of changes defended by many ecocentric theorists; I argue elsewhere,[4] how-ever, that it is neither necessary nor desirable that everyone live in decentralized, rural settlements and that there is a strong case to be made for the continued existence of urban settlements in an ecocentric society.) According to Parsons, human survival and well-being depend on a "knowl-edge and control of nature's substances and processes."[5] Yet the kind of knowledge Parsons and other orthodox eco-Marxists have in mind is not the kind that sees nature as a pattern or set of interrelationships to respect and follow or a design from which to draw guidance and inspiration. Rather, it is knowledge that will enable humans to overcome and redirect any resistance to their struggle for total mastery of nature. As Parsons explains,

> Marxism rejects this [ecocentric] unrealism, and in its view of the man–nature relation, it is inclined to emphasise man rather than the plants and the animals. If the assumption of the criticism is that Marxism has this emphasis, the assumption is correct. And like many modern humanisms, Marxism has sometimes overemphasized man's place in nature.[6]

Clearly, orthodox eco-Marxists regard ecocentrism as putting an unnecessary restraint on human development, which they regard as de-pendent upon an expanding science and technology that will increase our ability to control and manipulate the "secrets of nature." This perspective, of course, is entirely consistent with Marx's exclusive preoccupation with human betterment. Marx showed no interest in natural history, and he did not address the problem of nonhuman suffering. Indeed, Parsons defends Marx and Engels's lack of interest in the emerging "humane societies" for the prevention of cruelty to animals in the eighteenth and nineteenth centuries, arguing that the concern for the welfare of nonhuman animals was "a displacement of human concern" that was restricted by the privileged class position of its advocates.[7] In any event, Parsons suggests that in the long run nonhuman animals, like human animals, might also be liberated by technology:

> Presumably when animals are displaced entirely by machines as instruments of production, and when food is synthesized chemically, animals will enjoy a freedom not enjoyed since their domestication for food and labor in Neolithic times, and man's attitude toward them will likewise change with man's new freedom.[8]

However, far from being displaced by machines, many domestic animals (most notably cattle and hens) have been effectively *turned into* machines as a result of the application of "advanced" production techniques in agriculture. Moreover, ecological reality suggests that unless human population growth and the loss of genetic diversity and wild habitat is drastically curbed, there will be a very narrow range of nonhuman animals left to enjoy the distant "freedom" that Parsons believes will be wrought by advanced technology!

To the extent that environmental problems were acknowledged by Marx and Engels, they were attributed to the capitalist relations of production, not to the forces of production. Orthodox eco-Marxists have fully endorsed this analysis of the problem: the capitalist classes, while initially facilitating the development of the productive forces, are ultimately seen as acting as a fetter to their full development by standing in the way of a complete social appropriation of the power of nature and the control of any unwanted side effects. According to Parsons,

> an economic system [such as the capitalist one] that breaches the laws of nature by which wealth is produced will bring on inevitable reactions: impairment of nature's "metabolism" of ecological cycles, depletion of nature's resources, impoverishment of human society, *and a relapse of nature into the slumber of undevelopment* [my emphasis].[9]

The Marxist explanation for ecological degradation lies in the fact that capital works only for the benefit of the owners and controllers of capital rather than for the benefit of the complete society of producers. (It is assumed that the complete society of producers would act as a collective and would therefore be concerned to protect public environmental goods such as air, water, and soil.) It is thus the dynamic of *private* capital accumulation that has given rise to resource depletion, pollution, untrammeled urbanization, and the occupational and residential hazards suffered by workers and their families.[10] Yet, as John Clark observes, although Parsons tries to establish Marx's ecological credentials by arguing that Marx recognized an "essential incompatibility" between capitalism and "the system of nature," Parsons in fact misses Marx's antiecological point:

> Marx's point is not that this expansionism is in conflict with nature, but rather that capital's quest for surplus value contradicts and *limits* this development in some ways, to the detriment of humanity. In contrast, an ecological critique would question this very expansionism as being in contradiction with "the system of nature" [my emphasis].[11]

Orthodox eco-Marxists simply seek to replace the private and socially inequitable mastery of nature under capitalism with the public and socially equitable mastery of nature under communism. Ecological degradation under capitalism is seen by orthodox eco-Marxists as a measure of its *inefficiency,* of its failure to utilize resources wisely. As such, orthodox eco-Marxists are predominantly resource conservationists[12;] they are at home with Gifford Pinchot (with his "wise use" of natural resources argument), but are fundamentally at odds with John Muir's vision of large tracts of wilderness being protected in their "state of natural grace."[13] As we have seen, to the extent that orthodox eco-Marxists would be prepared to defend ecosystem preservationism, it would be on purely human-centered, instrumental grounds. Of course, Marx and Engels were also early pioneers of what I have called the "human welfare ecology" stream of environmentalism.[14] Indeed, Engels's classic critique of the working and living conditions of the Victorian working class is a major milestone in the development of this stream of environmentalism.[15] However, while Marx and Engels's critique challenged capitalism and exposed the misery of the working class, it did not challenge the hegemony of instrumental reason.

From an ecocentric perspective, the "true freedom" promised by scientific socialism is ultimately illusory. As Bookchins puts it, "At its best, Marx's work is an inherent self-deception that inadvertently absorbs the most questionable tenets of Enlightenment thought into its very sensibility."[16] Moreover, ecofeminists and social ecologists have drawn attention to the "masculine" character of the mastery sought by the mature Marx, who rejected as regressive the idolization of nature as a "nurturing mother." According to Marx, modern "man" must sever his umbilical cord with nature and become a self-determining being in order to achieve his "manhood." As John Clark has observed,

Marx's Promethean and Oedipal "man" is a being who is not at home
in nature, who does not see the Earth as the "household" of ecology.
He is an indomitable spirit who must subjugate nature in his quest for
self-realization. . . . For such a being, the forces of nature, whether in
the form of his own unmastered internal nature or the menacing powers
of external nature, must be subdued.[17]

Finally, orthodox eco-Marxists continue to place faith in the working
class as the agents of revolutionary change, both social and environmental.
According to Parsons, the proletariat still remain the class best situated for
"assuming a position of ultimate power and responsibility over the whole
transformed system"—including control of the unwanted side effects of the
manipulation of nature.[18] Post-Marxist green theorists, in contrast, have
challenged what they call the "productivist ideology" and inherent conservatism of the Western labor movement and have pointed instead to the
radical potential of new social movements, particularly those concerned
with ecology, feminism, and Third World solidarity.[19]

None of the above criticisms are intended to dismiss the eco-Marxist
critique of capitalism. Indeed, this critique still has relevance today in terms
of highlighting the ways in which capitalism can exploit laborer and land
alike, particularly in Third World countries where ecological considerations are displaced almost entirely in the drive to develop massive power
schemes and large-scale export industries (often merely to service the large
debts owed to First World countries).[20] From an ecocentric perspective,
however, the eco-Marxist critique simply does not go far enough; it is
fundamentally limited by its anthropocentrism, its focus on the relations
of production at the expense of the forces of production, and its uncritical
acceptance of industrial technology and instrumental reason.

Not surprisingly, the shortcomings of orthodox Marxism have attracted
the critical attention of a number of ecologically concerned Western
Marxists of a more "humanist" persuasion who are critical of "scientific
socialism" yet still attracted to Marxism as a philosophy and form of critique.

HUMANIST ECO-MARXISM

Unlike orthodox eco-Marxists, humanist eco-Marxists argue that it is
necessary to reassess Marx's technological optimism and his nineteenth-
century belief in material progress. According to André Gorz, the collective
appropriation by the proletariat of the capitalist forces of production would
not solve the ecological crisis: it would simply mean that the proletariat
would take over the machinery of domination.[21] Gorz has argued that the
development of the productive forces, hitherto welcomed by most Marx-

ists, must be reexamined on the grounds that they now threaten society's ecological support system.[22] Similarly, Enzensberger has argued that the ecological crisis can be dealt with in Marxist terms if we remember that capitalism is not just a property relation but also a *mode of production* in which the forces and relations of production are inextricably linked; this capitalist mode of production is something that the socialist countries, which he considers to be "still in transition," have yet to abandon.[23]

In asking how we might resolve the ecological contradictions of capitalism and what kind of human beings will inhabit the society that lies beyond the realm of domination, humanist eco-Marxists have sought inspiration from the philosophical writings of the young Marx.[24] Indeed, John Ely has shown that certain aspects of the young Marx's utopianism concerning the reconciliation of humanity and nonhuman nature have been taken up directly in the West German Greens' first economic program, although the usage of these concepts by the Greens is selective rather than systematic.[25]

The most ecologically sensitive case for a return to the ideas of the young Marx is that provided by Donald Lee in his essay entitled "On the Marxian View of the Relationship between Man and Nature."[26] However, although Lee's particular vision of humanist eco-Marxism is, for the most part, no longer vulnerable to the criticisms that I have leveled against orthodox eco-Marxism, a case can be made that it is also no longer *Marxist*. That is, Lee has developed a particular version of humanist eco-Marxism that downplays, and in some cases ignores, key distinctions and themes in the writings of the young Marx. Yet it is precisely these particular distinctions and themes (which, as we shall see, are endorsed by *other* versions of humanist eco-Marxism) that makes humanist eco-Marxism incompatible with an ecocentric perspective. The most important of these is Marx's distinction between "freedom" and "necessity." As we shall see, this key distinction serves to make the domination of the nonhuman world a *requirement* of human self-realization.

The basic goal of Lee's approach is to overcome alienation in a very broad sense, with the ecology crisis taken as evidence of our *alienation from nature* and as one more obstacle in the path to human emancipation. According to Lee's reading of Marx's early writings on alienation, there is no human–nature (or subject–object) dichotomy in Marx's thinking, as is so often claimed. Rather, Lee argues that Marx was concerned to overcome alienation between humans and themselves and their work and *between humans and external nature*. Under capitalism, both the worker *and* nonhuman nature are considered by the capitalist class as instruments to be exploited for private profit. According to Lee, "Capitalism was a necessary stage in man's development of the mastery of nature: but a further development is now necessary, namely, the overcoming of the dichotomy between man as subject over and against nature as object."[27]

In Lee's view, Marx's ideas were not anthropocentric because he conceived of nature as humanity's *inorganic body:* "This recognition of nature as our *body* will constitute the overcoming of the alienation of ourselves from nature, manifested in subject–object dualism."[28] On the basis of this insight, Lee has sought to outline an ecologically benign form of socialist stewardship of nature that will emancipate humans and nonhumans alike from the tyranny of capitalism so that humans can become the caretakers of their own "body." This socialist notion of nonhuman nature as *our* inorganic body, toward which we have a responsibility of care, is juxtaposed to the capitalist conception of nature as an alien "other" to be exploited for private profit. According to Lee, an ecological ethic must become part of the Marxist program of liberation; socialism must be developed to what Lee sees as its logical end, that is, beyond homocentrism (i.e., anthropocentrism).[29]

Lee argues that the present dichotomy between humanity and nature may be overcome only through the overthrow of the wasteful capitalist system (read "mode of production") and its replacement by a "rational, humane, environmentally unalienated social order."[30] The major features of this new order would be socially useful production, the reduction of labor time, maximum creative leisure, wise use of resources, rational population control, and solidarity between all living things, not just humans. Lee's postscarcity utopia is thus one in which everyone will be able to realize what Marx referred to as our "species being"–a situation of genuine freedom from need–whereby we can create according to the "laws of beauty" and become responsible stewards of the ecosystem. According to Lee, this form of socialist stewardship is based on a sense of enlightened self-interest because:

> Man is the *universal* being who can understand what is good for each species intrinsically, and thus, just as socialist man transcends the selfish greed of the capitalist and acts for the good of all men (which is ultimately his own good) so must ecologically aware socialist man transcend the selfish greed of homocentrism and act for the good of the whole ecosystem (which ultimately is his own good).[31]

Lee's socialist postscarcity utopia represents a considerable departure from the orthodox eco-Marxists' Promethean orientation toward nonhuman nature. Lee's socialist society does not seek to dominate nonhuman nature as an alien "other" through the development of technology. Rather, it is to be a society of self-determining individuals who realize themselves through free, conscious activity and who recognize nonhuman nature as but an extension of themselves, as part of their inorganic body.

However, what is *not* apparent in Lee's ecological interpretation of the ideas of the young Marx, and what *is* apparent in the writings of the young Marx and in other versions of humanist eco-Marxism, is a notion of human freedom that is irredeemably anthropocentric. This notion of human freedom takes its meaning from Marx's distinction between freedom and necessity.

It will be recalled that the young Marx maintained that humans realize their "essence" through unalientated labor and that the distinctive characteristic of humans was that "they produce universally, that is, produce even when free from physical need *and only truly produce in freedom thereof.*[32] This same distinction also runs through the writings of the mature Marx. For example, in *Capital* Marx wrote: "The realm of freedom begins only where labor which is determined by necessity and mundane considerations ceases; thus in the nature of things it lies beyond the sphere of actual material production."[33]

According to Marcuse and Gorz (both of whom build on this freedom–necessity distinction), the more we have mastered necessity, the more we can become truly free and realize our individuality through creative leisure, the sciences and the arts, convivial activity, and the like. According to Marcuse (who also drew on Freud's theory of human instincts), freedom lies in eros and play, not labor, for labor presupposes the suppression of instincts and the conquering of desire. In other words, the problem lies in the fact that social necessity demands that humans must always labor. Unlike Freud, however, Marcuse argued that in today's society scarcity (which gives rise to "basic repression") is not so much a brute fact as it is a consequence of a specific social organization that is sustained to secure the privileged position of powerful groups and individuals—a state of affairs that has led to "surplus" repression.[34] Marcuse observed that, paradoxically, the very technological achievements of "repressive civilization" (which he considered to be dominated by the "performance principle"—the prevailing historical form of the "reality principle") have created the preconditions for the gradual abolition of repression. That is, breaking down these social relations of domination would enable the forces of production to be pressed into the service of "genuine need" by liberating humans from toil, thereby creating a postscarcity and hence "nonrepressive civilization." Like the young Marx (and contra the mature Marx), Marcuse believed that both labor and scarcity could be abolished in this way rather than simply diminished.

The problem with this humanist eco-Marxist project of overcoming human alienation is that "true" human freedom and embeddedness in nature are posited as *inversely related.*[35] That is, the kind of freedom pursued by humanist eco-Marxists necessarily requires the subjugation of external nature (though labor's extension, technology) so that humans may ultimately become fully sovereign and answerable only to themselves, as opposed to being dependent on, and "held down" by, the limitations and inconveniences of nonhuman nature. Nonhuman nature remains, as Benton observes, "an external, threatening and constraining power . . . to be overcome in the course of a long-drawn-out historical process of collective transformation."[36]

The ultimate purport, then, of Marx's notion of "the resurrection of nature" in the *Economic and Philosophical Manuscripts of 1844* is not a nonanthropocentric socialist stewardship of "our inorganic body," as Lee would have us believe, but rather the further subjugation of the nonhuman world.[37] If Lee's humanist eco-Marxism is to remain recognizable as Marxism, then it must accept the anthropocentric implications of Marx's particular notion of freedom, which takes its meaning from the problematic freedom–necessity distinction.

In any event, some ecocentric critics have argued that even Lee's apparently benign ecocentric interpretation of the ideas of the young Marx—an interpretation that focuses on the theme of alienation—nonetheless serves to legitimize (albeit unwittingly) the domination of the nonhuman world. According to Lee, the overcoming of our alienation from nature is understood as the outcome of a dialectical struggle (sometimes referred to as a "metabolic interaction") between subject (the laborer) and object (external nature, the material to be transformed). According to this view, for nature to be recognized as our "body," a familiar and extended part of us rather than something "other," there must be a mutual interpenetration of both spheres through the activity of labor, resulting in the mutual transformation of both—thus arriving at the much-heralded "humanization of nature" and "naturalization of humanity." Ecocentric critics have suggested that this superficially attractive version of overcoming the human–nature dichotomy is yet another form of domination couched in the language of human self-realization.[38] According to Val Routley, Marx's early view of nature as our body, our creation, and our expression

> can usefully be seen as the product of Marx's well-known transposition
> of God's features and role in the Hegelian system of thought onto man.
> . . . Thus, Marx's theory represents an extreme form of the placing of
> man in the role previously attributed to God, a transposition so charac
> teristic of Enlightenment thought.[39]

The upshot of humanist eco-Marxism (Lee's version included) is that *the unity of humans with nature is achieved by making it our artefact, by totally domesticating it.* We have thus returned full circle to the orthodox solution to the environmental crisis (albeit couched in different language), to that stage of human development where nature is totally mastered through the power of associated individuals. In Lee's own words, this "unity" would enable us to live "in consciousness that each of us is identical with each other and with nature, and exploitation of men and nature would cease."[40]

From an ecocentric perspective, however, harmonizing our relationship with the rest of nature does not mean obliterating or humanizing what is "other " or not-human in nature. Rather, it means *identifying* with it (not making it identical to us) in a way that involves the recognition of the *relative autonomy* and unique mode of being of the myriad life forms that make up the nonhuman world. To ecocentric theorists, freedom or self-determination is recognized as a legitimate entitlement of *both* human and nonhuman life forms. The goal of an ecocentric political theory is "emancipation writ large"—the maximization of the freedom of *all* entities to unfold or develop in their own ways. Moreover, such freedom or self-determination is understood in relational terms (both socially and ecologically), in that the development of any relatively autonomous parts of a larger system (e.g., an ecosystem or the ecosphere) is inextricably tied to their relationships with the development of other relatively autonomous parts of that system as well as the development of the system itself (i.e., the whole). Ecocentric theorists argue that whereas the flourishing of human life and culture is quite compatible with a human life-style based on low material and energy throughput, the flourishing of nonhuman life *requires* such a human life-style. In order to meet this requirement, we need to live and experience ourselves as but one component of, and more or less keep pace with, the basic cycles and processes of nature, rather than seek to totally transcend the nonhuman world by removing all of its inconveniences and thereby obliterating its "otherness."

To be sure, Lee's reinterpretation of the "mastery of nature" to mean "rational harmony with nature" is more ecologically grounded than that of orthodox eco-Marxists, who would encourage the development of all manner of synthetic substitutes for "nature's bounty" so as to avoid remaining "tethered" to the cycles of nature. Yet this latter kind of outcome is logically entailed in the quest for freedom according to the young Marx. As we have seen, if true freedom is understood to be inversely related to our embeddedness in nature, then the realization of that freedom necessarily requires us to increase our control over, and reduce our dependence on, ecological cycles. The upshot is that nature, although redefined as "our body," must be thoroughly tamed and made subservient to human ends.

Of course, the kind of "true human freedom" promised by humanist eco-Marxists is an attractive and familiar interpretation of "freedom," made all the more so in the light of the general bifurcation between work and leisure in modern society. Yet, from an ecocentric perspective, this eco-Marxist coupling of necessity–freedom and work–leisure is objectionable in two important respects. First, it is based on the anthropocentric Marxist "differential imperative" of *homo faber*. Arguing by way of the "differential imperative" means selecting certain characteristics that are believed to be special to humans vis-à-vis other species as the measure of both human virtue and human superiority over other species. According to Marx, to be fully human and truly free, humans must maximize what he believed made us different from the rest of nature, namely, our ability to self-consciously act upon and transform the external world and thereby augment our own powers.[41] Yet, like so many anthropocentric assumptions, Marx's putative human–nonhuman opposition is based on an erroneous understanding of ecological reality. As Benton puts it in an extended critique of this opposition,

> For his intellectual purposes, Marx exaggerates both the fixity and limitedness of scope of the activity of other animals, and the flexibility and universality of scope of human activity upon the environment.[42]

Second, the eco-Marxist distinction between freedom and necessity reifies nonessential activities as the means to true individual fulfillment while downgrading as lowly or "animal-like" many life-sustaining activities that can be potentially more fulfilling if approached and organized differently. (By "life-sustaining activities," I am referring to those fundamental human activities necessary for survival and physical and psychological health, such as growing and preparing food, constructing and maintaining shelter, nurturing and teaching the young, and caring for the infirm and the elderly.) The result is that culture and self-expression are made the complete antithesis of necessary material labor. This is because the general reduction of necessary labor time is seen to provide the foundation for the "true freedom" to be experienced by all humans, a freedom to be "purchased" via ongoing technological developments designed to "relieve" humans from concerning themselves with those burdensome tasks that have limited the development of present and prior generations of humans and which are seen to remain forever the lot of the rest of the animal world.[43]

Marx's distinction between freedom and necessity thus creates a dualism not only between humans and nonhuman nature but also *within* human nature between common, lowly animal functions, powers, and needs and sui generis, higher human functions, powers, and needs.[44]

Simone de Beauvoir has drawn attention to a similar kind of contrasting valuation between nature and culture and between the self-limiting work of women and the self-transcending work of men. The former is treated as private, mundane, and concerned with the regeneration and repetition of life, while the latter is regarded as public, worthy, and concerned with transcending life by reshaping the future through technology and symbols.

The ecocentric objection to the postscarcity utopia of humanist eco-Marxism is that it would cultivate a type of human who, as Val Routley has observed, through science and technology, is thoroughly insulated from, and in control of, the cycles of nature and the myriad of other nonhuman life forms. Indeed, it is hard to see how the overcoming of human alienation from nature is to be achieved in such a utopia, when humans are to be so thoroughly insulated and removed from their biological roots. In effect, the price of overcoming alienation in the workplace is alienation from nonhuman nature. Moreover, as Bookchin has argued, according to the Marxist view of freedom, class society and authoritarian social relations will remain unavoidable for so long as the mode of production fails to provide a sufficient material abundance for everybody to enjoy the realm of "true freedom." Until that time, the "realm of necessary" will become a "realm of command and obedience, of ruler and ruled."[45] That is, the domination of people and the rest of nature under a rationalized capitalism remains the precondition for the achievement of a distant and continually postponed socialist freedom.

It is clear that Marcuse's version of humanist eco-Marxism is more firmly grounded in the philosophical ideas of the young Marx than Lee's version. Indeed, it is noteworthy that Tolman has rejected Lee's selective reading of Marx's early writings for being decidedly un-Marxist. According to Tolman, Lee's basic argument that we ought to see ourselves as stewards of the environment, which is to be seen as part of our "inorganic" body, is a serious distortion of Marx's true position, since Marx ultimately rejected the "early ideas" drawn upon by Lee as abstract and idealist.[46] We have already seen that the mature Marx had come to the view that the antagonistic struggle between humanity and nature would never be completely resolved, that is, nature would never be fully "resurrected" since labor could never be totally abolished and, accordingly, there would always be some parts of nature that remained untransformed and alien. For his part, Tolman makes no apologies in declaring the incompatibility between Marxism and ecocentrism. On this score, Tolman seems to have a clearer perception than Lee of the long-term technological and ecological implications of Marxism.

From an ecocentric perspective, then, the Marxist dichotomy between freedom and necessity must be transcended if we are to allow the mutual unfolding of both human and nonhuman life. In particular, the view (strongly endorsed by Gorz in particular) that even unalienated, self-managed material labor is a "lower" form of freedom than unnecessary labor and/or leisure activity must be rejected. Rather, the emphasis must turn to exploring the many ways in which basic needs may be met and necessary and life-sustaining work performed in a manner that is personally, aesthetically, and intellectually satisfying and *not* environmentally damaging.[47] Individuality, self-expression, and rounded human development may then be realized *through* socially useful work as well as through other kinds of activity (one kind of setting in which rounded human development is possible is a self-managed, cooperative community). Moreover, contrary to Gorz's presumption, the enjoyment of "true freedom" as leisure time need not be dependent on high technology or a high energy and material throughput. As Marshall Sahlins has argued in *Stone Age Economics*, the enjoyment of affluence (interpreted here as ample creative leisure rather than an abundance of material goods) by the members of a particular society is not necessarily dependent on that society "mastering necessity" by conquering nature through advanced technology and a high energy consumption.[48] Quite the contrary, his book is an important illustration of the maxim "Want not, lack not" and a challenge to what he calls modern culture's "shrine to the Unattainable: Infinite Needs."[49] From an ecocentric perspective, such creative leisure is best procured through the critical revision and simplification of human needs and the development of tools and goods appropriate to those revised needs than through the systematic replacement of human labor by energy-intensive machines.

BEYOND MARXISM

So far I have sought to show that an ecocentric perspective cannot be wrested out of Marxism, whether orthodox or humanist, without seriously distorting Marx's own theoretical concepts. As John Clark has put it, "To develop the submerged ecological dimensions of Marx would mean the negation of key aspects of his philosophy of history, his theory of human nature, and his view of social transformation."[50] This explains why nonanthropocentric green theorists have chosen not to develop their ideas within a Marxist framework and instead have sought guidance from other traditions of political thought such as utopian socialism, communal anarchism, and feminism. Far from providing a theoretical touchstone, Issac Balbus has argued that the time has come for ecologically oriented political theorists to do to the ideas of Marx what he did to the ideas of the bourgeois thinkers he contested, namely, explain their origins in order to reveal their historical limits.[51] In this respect, Hwa Yol Jung has provided three succinct reasons why ecologically oriented theorists should abandon the ideas of Marx:

> First, he was too Hegelian to realize that the gain of "History" (or Humanity) is the loss of "Nature"; second, he was influenced by the

English classical labor theory of value which undergirds his conception
of man as *homo faber;* and third, he was a victim of the untamed optimism
of the Enlightenment for Humanity's future progress.[52]

To be sure, Marx's conception of freedom was more comprehensive
than the liberal concept of freedom that he called to account. As Booth
has neatly put it, whereas liberalism had grasped "one form of unfreedom,
coercion, or the arbitrary rule of one will over another, which was the
dominant form in pre-capitalist societies," Marxism recognized another
form, namely, the silent and objective compulsion of the economic laws of
capitalism.[53] As ecocentric theorists have shown, however, neither liberalism nor Marxism has acknowledged the unfreedom of the nonhuman
world under *industrialism.*[54] To acknowledge the particular kinds of unfreedom exposed by Marx, as ecocentric theorists do (given their general
concern for "emancipation writ large") does not also require an acceptance
of Marx's notion of freedom. Quite the contrary, I have sought to show
that Marx's notion of freedom as *mastery* achieved through struggle; as the
subjugation of the external world through labor and its extension, technology; as the conquering of mysterious or hostile forces and the overcoming
of all constraints, can be achieved only at the expense of the nonhuman
world.

Extract 2, from Eckersley's Chapter 6: "Ecosocialism: The Post-Marxist Synthesis"

INTRODUCTION

The late Raymond Williams once described the ecology movement as "the
strongest organised hesitation before socialism."[55] Ecosocialism—a position
Williams himself defended in his later writings—represents a concerted
attempt to revise and reformulate the democratic socialist case in the light
of the ecological challenge. Ecosocialists have also used this opportunity for
theoretical stocktaking to respond to other significant challenges facing
socialism—challenges that form part of, but are not unique to, the ecological
critique—in an effort to address the concerns of new social movements and
to recapture the "high ground" of emancipatory discourse. As Frieder Otto
Wolf has put it, "A socialism without qualification will never again be able
to become a hegemonic force within emancipatory mass movements."[56]

The three most significant challenges facing socialism that ecosocialists have sought to address are (1) the historical legacies of bureaucratization, centralization, and authoritarianism; (2) the problematic role of the working class as the agent of revolutionary change; and (3) the disillusionment with the traditional "productivist" and growth-oriented socialist response to the indignities of poverty, which has usually been to augment the economic power of the state, seek a better mastery of nature through modern scientific techniques, and step up production. Ecosocialists have sought to respond to these historical legacies by reasserting the principles of democratic self-management and production for human need. According to Williams, "This is now our crisis: that we have to find ways of self-managing not just a single enterprise or community but a society."[57]

The ecosocialist theory presented in this chapter represents the most influential family of socialist thought in green circles. It has emerged from a critical dialogue between Marxist orthodoxy, various currents of Western Marxism, and Western social democracy, on the one hand, and the radical environmental movement, on the other hand. The resulting body of theory may be described as largely post-Marxist insofar as it is highly critical of orthodox Marxism (and much Western Marxism), but it is not *anti*-Marxist. Many theorists within this tradition occasionally draw on Western Marxist insights alongside other older traditions and contemporary strands of socialist thought, including utopian socialism, the self-management ideas of the New Left, and socialist feminism. While ecosocialists also share the anticapitalist and self-management orientation of ecoanarchists, they generally argue (contra ecoanarchists) that the state must play a key role in facilitating the shift toward a more egalitarian, conserver society. . . .

Ecosocialists argue that the logic of capital accumulation is fundamentally incompatible with ecological sustainability and social justice. Accordingly, they argue that capitalism must be largely replaced with a nonmarket allocative system that ensures ecologically benign production for genuine human needs. The real challenge facing ecosocialists, however, is how to develop new, democratic, and noncentralist social institutions that are able to give expression to ecosocialist values such as self-management, producer democracy, and the protection of civil and political liberties.

The major thematic innovations of ecosocialism—the rejection of the economic growth consensus, the emphasis on ecologically benign production for human need, the attempt to widen the productivist outlook of the labor movement and encourage a critical dialogue between the labor movement and new social movements, and the new internationalism—together represent a major overhaul of socialist thought. Moreover, these theoretical revisions place ecosocialism squarely within the spectrum of green or emancipatory political thought.

However, as we shall see, ecosocialism has self-consciously declined to step across the anthropocentric divide and embrace an ecocentric perspective. Instead, most ecosocialists have rejected the need for a "new ecological paradigm" and have argued that socialist thought provides a sufficient repository of values for ecological and social reconstruction.

THE MEANING AND LESSON OF ECOLOGY
ACCORDING TO ECOSOCIALISM

The heart of the philosophical difference between ecocentrism and ecosocialism concerns the meaning and relevance of ecology to emancipatory theory. Ecosocialists regard the demands of the environmental movement for a safe and healthy environment as but a subset of the modern radical project. In particular, we have seen that many ecosocialists regard the radical environmental movement (by this they mostly have in mind what I have characterized as the "human welfare ecology"[58] stream) as part of a larger struggle to *overcome capitalism.* The radical environmental movement is seen to be part of that larger struggle because it highlights the incompatibility of market rationality with ecological limits by revealing the many ways in which the economic "externalities" of private capital have seriously compromised human welfare, health, and survival.[59]

What is at stake, and what is now attainable, according to ecosocialism, is the full realization of *human autonomy* within a safe and healthy physical environment and a democratic and cooperative social environment. Significantly, most ecosocialists reject the idea that ecology can effect a fundamental paradigm shift in political theory along the lines suggested by many green theorists.[60] Ecosocialists argue that a preoccupation with ecological principles leads to an excessive preoccupation with "nature protection" and deflects attentions away from the *social* origins of environmental degradation.[61] What must be grasped, they argue, is that the "environment" is an essentially *human* context that is *socially* determined rather than something before which we humbly "submit."[62] More generally, ecosocialists argue that if we wish to retain a commitment to the modern political ideas of justice, equality, and liberty, then we must look to the lessons of human history (as interpreted by the various tributaries of socialist thought) rather than to natural history. Indeed, Gorz has gone so far as to declare that it is "impossible to derive an ethic from ecology."[63]

In support of the argument that ecological principles cannot provide the basis for a new politics, ecosocialist theorists frequently point out that it is possible to have a society that respects ecological limits but is undemocratic and authoritarian.[64] As we have seen, ecosocialists reject the claimed "newness" of the green movement and the idea that it has transcended old political rivalries and instead point out the continuities between green politics and many strands of socialism.[65] The only "newness" of the green movement is seen to reside in its recognition of "ecological limits"—something that ecosocialists agree cannot be ignored.[66] "The point," argues Gorz, "is not to defy nature or to 'go back' to it, but to take account of a simple fact: human activity finds in the natural world its external limits."[67]

Yet the ecosocialist argument that it is impossible to "derive" an ethic from ecology is misleading and creates an overdrawn opposition: that ecocentrism represents a naive form of authoritarian ecological determinism, while ecosocialism recognizes the active presence of humankind in constructing and shaping the environment. The environmental ethics of ecosocialism and ecocentrism are both *informed* by ecological insights, but the environmental ethic of ecosocialism is simply a different and more limited kind of environmental ethic than that of ecocentrism. That is, the ecosocialist ethic is a prudential ethic that largely represents an amalgamation of the resource conservation and human welfare ecology perspectives, both of which are informed by the life sciences such as ecology but which ultimately rest on anthropocentric *norms* of human autonomy, health, and welfare. Ecocentrism is also informed by the life sciences, but it, too, finds its ultimate justification in a normative rather than in a scientific framework. (As I've argued, to appeal to nature as known by the science of ecology rather than to ethics as the ultimate arbiter of a green political theory is misguided and does not in itself amount to a justification for a particular political posture.[68]) This ecocentric normative framework subsumes the human-centered ecosocialists' norms of autonomy, health, and welfare in a broader ecological framework that seeks the mutual flourishing of *all* life forms. Such a perspective does not seek to downgrade human creativity nor deny the extent to which humans influence ecological and evolutionary processes. Rather, it asks that we employ our creativity to develop technologies and life-styles that allow for the continuation of a rich and diverse human *and* nonhuman world. Ecosocialism, in contrast, may be seen as merely fusing human welfare ecology with democratic socialism, but transcending neither.

To return to the ecosocialist critique, if the only "lesson" provided by ecology is one of physical limits to growth, then it is indeed possible to have a range of different political regimes—including authoritarian and fascist ones—that observe such limits.[69] Robert Heilbroner's *An Inquiry into the Human Prospect*, which is essentially concerned with human survival, is a clear case in point.[70] Of course, an authoritarian regime might be successful—at least in the short term—in ensuring human survival (or, more likely, the survival of certain privileged classes of humans) and quite possibly the (indirect) survival of many nonhuman life forms. However, it would achieve this by severely restricting opportunities for democratic participation and self-determination, a route that is incompatible with the general ecocentric norm of mutual unfolding of *both* the human and the nonhuman worlds. For ecosocialists to reject ecocentrism on the ground that it does not rule out fascism is to miss the *inclusive* nature of the ecocentric norm of "emancipation writ large."

The ecosocialist rejection of the idea of a paradigm shift in green political theory is generally correct insofar as it applies to interhuman struggles. When viewed from the perspective of the traditional political spectrum, ecocentrism is, and must continue to be, generally "more left than right" in contending with old political rivalries based on differentials in wealth, power, and social privilege. However, ecocentric theory is most certainly new in the way it seeks to reorient humanity's relationship to the rest of nature. In this respect, it represents a new constellation of ideas that challenges the anthropocentric assumptions of post-Enlightenment political thought and calls for a more radical reassessment of human needs, technologies, and life-styles than ecosocialism.

To be sure, we have seen that ecosocialism has itself traveled some distance down this new path insofar as it has acknowledged the many ways in which capitalism objectifies and commodifies both people *and* nonhuman nature.[71] However, the ecosocialist critique of instrumental reason, like that of the Frankfurt School, ultimately comes to rest on the human-centered argument that it is wrong to dominate nature because it gives rise to the domination of people. For example, Gorz, in noting that the disregard of ecological limits will often set off an unwelcome ecological backlash, argues that

> it is better to leave nature to work itself out than to seek to correct it at the cost of a growing submission of individuals to institutions, to the domination of others. For the ecologist's objection to system engineering is not that it violates nature (which is not sacred), but that it substitutes new forms of domination for existing natural processes.[72]

Of course, this ecosocialist concern for human autonomy is laudable in and of itself. From an ecocentric perspective, however, it means that the case for the recognition and protection of nonhuman species is activated only when it can be shown to facilitate human emancipation. As Rodman has observed, Gorz has an intuition that we should "let nature be" not because it is sacred or has its own relative autonomy but because it makes *us* freer.[73] While such an argument has a place within ecocentric theory (indeed, it serves to bolster ecocentric theory in that it shows that the flourishing of human *and* nonhuman life need not be a zero-sum game), it provides no defense for threatened nonhuman species in those cases where there is no appreciable link with human domination and where such species appear to provide no present or potential use or interest to humankind. Moreover, such an argument also serves to reinforce anthro-pocentric attitudes. As John Livingston has aptly put it, at best, wildlife might "emerge as a second-generation beneficiary" from human welfare ecology reforms.[74] In this respect, Raymond Williams's views on wildlife preservation are telling:

> We are not going to be the people . . . who simply say "keep this piece clear, keep this threatened species alive, at all costs." The case of a threatened species is a good general illustration. You can have a kind of animal which is damaging to local cultivation, and then you have the sort of problem that occurs again and again in environmental issues. You will get the eminences of the world flying in and saying: "must save this beautiful wild creature." That it may kill the occasional villager, that it tramples their crops, is unfortunate. But it is a beautiful creature and it must be saved. Such people are the friends of nobody, and to think that they are allies in the ecological movement is an extraordinary delusion.[75]

This, of course, is consistent with my identification of wilderness or wildlife preservation as one of the "litmus tests" that enables us to distin-guish ecocentric from anthropocentric green theorists. That is, wherever there is an apparent conflict between human interests and the interests of nonhuman species (in this case the protection of wildlife that appear to be of no use to humankind), ecosocialists *consistently dismiss* nonhuman interests.

Similarly, to the extent that ecosocialists have contributed to the human population debate (the other litmus test), it is usually by way of a critique of what are seen as the "neo-Malthusian" arguments of population control advocates such as Paul Ehrlich—a critique that follows the spirit, if not the letter, of Marx's critique of Malthus.[76] According to this argument, the real causes of resource scarcity, famine, and environmental degrada-tion are not the existence of too many people or the limited carrying capacity of the Earth but rather social factors such as the maldistribution of resources and inappropriate technology, which arise under the capitalist mode of production.[77] The ecosocialist solution, then, is not population control but the replacement of capitalism with a cooperative social order that uses ecologically appropriate technologies for the satisfaction of human need.[78]

From an ecocentric perspective, the ecosocialist response goes only part of the way toward addressing the population problem. First, it fails to consider the many ways in which growing absolute numbers of humans can magnify environmental degradation and therefore impair the overall qual-ity of human life. Second, it fails to consider the impact of growing absolute numbers of humans on the nonhuman community—a limitation that arises from the exclusive ecosocialist preoccupation with human welfare. The environmental impact of humans is a function not only of technology and affluence (i.e., level of consumption), but also absolute numbers of humans.[79] From an ecocentric perspective, it is not enough simply to wait for the "demographic transition" (i.e., the lower birth and death rates that usually follow improved living standards) to achieve a stable and well-fed human population, because the price of such a transition is further widespread ecological degradation and species extinction. To minimize ecological degradation during this transition period, ecocentric theorists argue that it is necessary to bring about, *in addition to* technological and distributional reforms and a lowering of resource consumption, a wide range of humane family planning measures with a view to stabilizing and then reducing human population. "Humane family planning measures" include free contraceptives and free birth control information and counseling; affirmative action to improve the status and social opportunities of women; and ecological education campaigns that explain, inter alia, the impact of human population growth on ecosystems and the need to reduce the size of families to one or two children. The synergetic effect of introducing ecologically benign technologies and lowering energy and resource consumption *as well as* lowering the birth rate would have a much more dramatic result in terms of lessening environmental degradation and protecting biotic diversity than would the more limited ecosocialist solution.

The foregoing critique of the anthropocentric premises of ecosocialist thought does not require a rejection of either the entirely defensible socialist concern to find an allocative system that ensures production for genuine human need *or* the more general and equally defensible concern to seek the mutual self-realization of all humans. Quite the contrary, both of these concerns fall naturally into the orbit of the ecocentric perspective defended in this inquiry. Indeed, there is already a strong resonance between ecocentric social goals and key ecosocialist goals such as the new internationalism, democratic participation, and ecologically sustainable production for human need. Moreover, both ecocentrism and ecosocialism reject an atomistic model of reality in favor of a reciprocal model of internal relations (albeit with different ethical horizons and implications).[80] These responses open up the possibility of theoretical bridge building between ecocentrism and ecosocialism. That is, many ecosocialist principles and arguments can be selectively incorporated into the broader theoretical framework of ecocentrism *once they are divested of their anthropocentric limitations.*

The upshot of such theoretical bridge building for ecosocialism would be a widening of its field of moral considerability so that it reaches beyond the human community to include all of the myriad life forms in the biotic community. In particular, this would mean a broadening of the ecosocialist approach to wilderness protection and human population growth in accordance with ecocentric goals. More generally, it would mean a broadening of the context of political, economic, and technological decision making so that human interests are pursued, wherever practicable, in ways that *also* enable other life forms to flourish.

The upshot for ecocentrism would be a strengthening and broadening of its political and economic analysis that would make it better equipped to determine the kinds of institutional changes and redistributive measures that would be required to ensure an equitable transition toward a sustainable and more cooperative society. It would also enable ecocentrism to anticipate and address in a more concerted way the various forms of opposition that are likely to be encountered in the attempt to give practical expression to ecocentric emancipatory goals. In this respect, ecosocialists are right to argue that capital will not be placed at the service of emancipatory goals without increasing government intervention in the market and without a gradual democratization of the economy.

NOTES

1. See Charles Tolman, "Karl Marx, Alienation, and the Mastery of Nature," *Environmental Ethics,* 3, 1991, 63–74; Howard Parsons, *Marx and Engels on Ecology* (Westport, Conn.: Greenwood Press, 1978); and Melanie Beresford, "Doomsayers and Eco-nuts," *Politics,* 12, 1991, 98–106.

2. Tolman, "Karl Marx," 73.

3. Parsons, *Marx and Engels,* 44.

4. In Robyn Eckersley, *Environmentalism and Political Theory: Toward an Ecocentric Approach* (Albany: State University of New York Press, 1992), chap 7.

5. Parsons, *Marx and Engels,* 45.

6. Ibid.

7. Ibid., 47.

8. Ibid., 45.

9. Ibid., 16.

10. Ibid., 18–19.

11. John Clark, "Marx's Inorganic Body," *Environmental Ethics,* 11, 1984, 245.

12. See the Introduction to Part IV of this book.

13. For a juxtaposition of the respective concerns and philosophies of Marx and Muir (and their successors), see Frances Moore Lappé and J. Baird Callicott, "Marx Meets Muir: Towards a Synthesis of the Progressive Political and Ecological Visions," *Tikkun,* 2(3), 1987, 16–21; and Robyn Eckersley, "The Road to Ecotopia: Socialism versus Environmentalism," *Island Magazine,* Spring, 1987, 18–25; reprinted in *Ecologist,* 18, 1988, 142–147; *Trumpter,* 5, 1988, 60–64; and in *First Rights: A Decade of Island Magazine,* eds. Andrew Sant and Michael Denholm (Elwood, Victoria: Greenhouse, 1989), 50–60.

14. See the Introduction to Part IV of this book.

15. Friedrich Engels, *The Condition of the Working Class in England,* 2d ed., trans. and ed. W. O. Henderson and W. H. Chaloner (Oxford: Basil Blackwell, 1971).

16. Murray Bookchin, "Marxism as Bourgeois Sociology," in *Toward an Ecological Society* (Montreal: Black Rose, 1983), 195.

17. Clark, "Marx's Inorganic Body," 256.

18. Parsons, *Marx and Engels,* 14.

19. See, e.g., Bookchin, *Toward an Ecological Society;* Rudolph Bahro, *Socialism and Survival* (London: Heretic Books, 1982); John Clark, *The Anarchist Moment: Reflections on Culture, Nature, and Power* (Montreal: Black Rose, 1984); and Carl Boggs, *Social Movements and Political Power: Emerging Forms of Radicalism* (Philadelphia: Temple University Press, 1986).

20. See Michael Redclift, *Development and the Environmental Crisis: Red or Green Alternatives?* (London: Methuen, 1984), and *Sustainable Development: Exploring the Contradictions* (London: Methuen, 1987).

21. André Gorz, *Farewell to the Working Class: An Essay in Post-Industrial Socialism,* trans. M. Sonenscher (London: Pluto, 1982), 100.

22. André Gorz, *Ecology as Politics,* trans. Patsy Vigderman and Jonathan Cloud (London: Pluto, 1980), chap. 1.

23. H. M. Enzensberger, "A Critique of Political Ecology," *New Left Review, 84,* 1974, 21; reprinted as Chapter 1, this volume. Enzensberger's call has also been endorsed by Adrienne Farago, "Environmentalism and the Left," *Urban Policy and Research, 3,* 1985, 13.

24. See, e.g., Donald Lee, "On the Marxian View of the Relationship between Man and Nature," *Environmental Ethics, 2,* 1980, 3–16; Herbert Marcuse, *One-Dimensional Man* (London: Routledge and Kegan Paul, 1964) and *Counterrevolution and Revolt* (London: Allen Lane, 1972)—see in particular the chapter "Nature and Revolution," esp. 63–64; K. D. Shifferd, "Karl Marx and the Environment," *Journal of Environmental Education, 3,* 1972, 39–42; André Gorz, *Ecology as Politics* and *Farewell to the Working Class;* Janna Thompson, "The Death of a Contradiction," *Intervention, 17,* 1983, 20; and Michael Lowy, "The Romantic and the Marxist Critique of Modern Civilization," *Theory and Society, 16,* 1987, 891–904.

25. John Ely, "Marxism and Green Politics in West Germany," *Thesis Eleven, 13,* 1986, 26.

26. Lee, "Marxian View." It should be noted that Lee's case had already been advanced as early as 1972 by K. D. Shifferd, in "Karl Marx," and by Herbert Marcuse in his essay "Nature and Revolution" in *Counterrevolution and Revolt.* However, Lee's argument is a more suitable focus for my present purposes since it is both more recent and more developed.

27. Lee, "Marxian View," 11.

28. Ibid., 8.

29. Ibid., 15.

30. Ibid., 11.

31. Ibid., 16.

32. Karl Marx, *Economic and Philosophical Manuscripts of 1844,* trans. M. Milligan, ed. D. J. Struik (New York: International, 1964), 113.

33. Marx, *Capital* (London: Lawrence and Wishart, 1973) 3:820. Against passages of this kind, Schmidt has noted passages from the *Grundrisse* (which offers an important bridge between the young and the mature Marx) that suggest that Marx believed that a humanized realm of necessity can also become a sphere of human self-realization (*The Concept of Nature in Marx* [London: New Left Books, 1971], 143). This does not, however, detract from Marx's overriding concern to rationalize and reduce necessary labor. As Schmidt himself argues, "The problem of human freedom is reduced by Marx to the problem of free time" (142). This was also a concern of young Marx and certainly represents the direction in which Gorz and Marcuse have developed Marx's ideas.

34. According to Marcuse, Freud had theorized that "behind the reality principle lies the fundamental fact of Ananke or scarcity (*Lebensnot*), which means that the struggle for existence takes place in a world too poor for the satisfaction of human needs without constant restraint, renunciation, delay." See his *Eros and Civilisation: A Philosophical Inquiry into Freud* (London: Routledge and Kegan Paul, 1956), 35.

35. Isaac D. Balbus, *Marxism and Domination* (Princeton, N.J.: Princeton University Press, 1982), 274.

36. Ted Benton, "Humanism vs. Speciesism? Marx on Humans and Animals," *Radical Philosophy*, Autumn, 1988, 7.

37. Balbus, *Marxism and Domination*, 274.

38. Val Routley, "On Karl Marx as an Environmental Hero," *Environmental Ethics, 3*, 1981, 239.

39. Ibid., 239–240.

40. Lee, "Marxian View," 9.

41. Benton, "Humanism vs. Speciesism," 8.

42. Ibid., 9.

43. There are passages in the *Grundrisse* that run contrary to this interpretation in that they suggest that human freedom or self-realization can be attained *through* democratic self-management in the workplace. This is a much more defensible interpretation from an ecocentric perspective. However, I argue that this interpretation is inconsistent with Marx's freedom–necessity distinction, which appears in both his early and mature writings.

44. Benton, "Humanism vs. Speciesism," 12.

45. Bookchin, "Marxism," 204–206. See also Routley, "On Karl Marx," 241.

46. Tolman, "Karl Marx," 72.

47. See also Benton, "Humanism vs. Speciesism," 14; Routley, "On Karl Marx," 242.

48. Marshall Sahlins, *Stone Age Economics* (London: Tavistock, 1974), esp. chap. 1, "The Original Affluent Society."

49. Ibid., 39.

50. Clark, "Marx's Inorganic Body," 250.

51. Isaac D. Balbus, "A Neo-Hegelian, Feminist, Psychoanalytical Perspective on Ecology," *Telos, 52*, 1982, 140.

52. Hwa Yol Jung, "Marxism, Ecology, and Technology," *Environmental Ethics, 5*, 1983, 170.

53. William James Booth, "Gone Fishing: Making Sense of Marx's Concept of Communism," *Political Theory, 17,* 1989, 220.

54. As I pointed out in chapter 1 of *Environmentalism and Political Theory,* Mill and Bentham constitute two importance exceptions within the liberal tradition insofar as Mill entertained the idea of a stationary state economy and Bentham considered all sentient beings to be morally considerable.

55. Raymond Williams, "Hesitations before Socialism," *New Socialist,* September, 1986, 35–36.

56. Frieder Otto Wolf, "Eco-Socialist Transition on the Threshold of the Twenty-First Century," *New Left Review, 158,* 1986, 35.

57. Williams, "Hesitations," 34.

58. See the Introduction to Part IV of this book.

59. David Pepper, "Radical Environmentalism and the Labour Movement," in *Red and Green: The New Politics of the Environment,* ed. Joe Weston (London: Pluto, 1986), 115. See also Pepper, *The Roots of Modern Environmentalism* (London: Croom Helm, 1984), 199.

60. See, e.g., Martin Ryle, *Ecology and Socialism* (London: Century Hutchinson, 1988), 7–8; Weston, "Introduction" to *Red and Green,* 2; and Pepper, "Radical Environmentalism," 121. See also Pepper, "Determinism, Idealism, and the Politics of Environmentalism—A Viewpoint," *International Journal of Environmental Studies, 26,* 1985, 11–19.

61. Indeed, Pepper describes such concerns as reactionary and "largely an elitist defence of what a minority of ex-urbanites saw as 'wild nature' or 'traditional landscapes' " (Pepper, "Radical Environmentalism," 121fn5). See also Frankie Ashton, *Green Dreams, Red Realities* (N.A.T.T.A. Discussion Paper No. 2, Alternative Technology Group, The Open University, Milton Keynes, U.K., 1985).

62. See, e.g., Weston, *Red and Green;* Pepper, "Radical Environmentalism."

63. André Gorz, *Ecology as Politics* (London: Pluto, 1980), 16.

64. Ibid., 17.

65. David Pepper, *Roots of Modern Environmentalism,* 193–194. See also Peter C. Gould, *Early Green Politics: Back to Nature, Back to the Land, and Socialism in Great Britain, 1880–1900* (Brighton, U.K.: Harvester, 1988).

66. The terms of this debate were framed as early as 1974 by Hans Magnus Enzensberger in "A Critique of Political Ecology," *New Left Review, 84,* 1974, 23–31. Since then, most of the discussion of the ecological crisis by ecosocialist theorists (e.g., Ryle, Gorz, Weston, Pepper, Bell, and Hulsberg) has been mainly couched in the language of "ecological limits" or "constraints" on human action.

67. Gorz, *Ecology as Politics,* 13.

68. See Eckersley, *Environmentalism and Political Theory,* chap. 3.

69. For a discussion of the ecological ideas in Nazism, see Anna Bramwell, *Ecology in the 20th Century: A History* (Cambridge: Cambridge University Press, 1989).

70. Robert L. Heilbroner, *An Inquiry into the Human Prospect* (New York: Norton, 1974).

71. See, e.g., Raymond Williams, *Towards 2000* (Harmondsworth, U.K.: Penguin, 1983), 214–215.

72. Gorz, *Ecology as Politics,* 18.

73. John Rodman, review of *Ecology as Politics* by André Gorz, *Human Ecology*, *12*, 1984, 324.

74. John Livingston, *The Fallacy of Wildlife Conservation* (Toronto: McClelland & Stewart, 1982), 42.

75. Raymond Williams, *Socialism and Ecology* (London: Socialist Environment and Resources Association, n.d.), 14. See also Williams, *The Country and the City* (London: Chatto and Windus, 1973), 82.

76. See, e.g., Enzensberger, "Critique," 13–15. Marx has argued that the apparent phenomenon of overpopulation under capitalism arose not as a result of natural conditions but rather as a result of the contradictions in the capitalist relations of production—in particular, its need to maintain an "industrial reserve army." For a discussion, see Michael Perelman, "Marx, Malthus, and the Concept of Natural Resource Scarcity," *Antipode, 11*, 1979, 80–84.

77. These arguments received a considerable public airing in the debate between Barry Commoner and Paul Ehrlich in the early 1970s. On the Ehrlich–Commoner debate, see Paul Ehrlich, *The Population Bomb*, rev. ed. (London: Pan/Ballantine, 1972), and Barry Commoner, *The Closing Circle: Nature, Man and Technology* (New York: Bantam, 1972). For an exchange of views, see Paul Ehrlich, John Holdren, and Barry Commoner, "Dispute: *The Closing Circle*," *Environment*, *14*, 1972, 24–24, 40–52.

78. See, e.g., Pepper, *Roots of Modern Environmentalism*, 167–169. On the more specific problem of world hunger, most ecosocialists focus on the need for land redistribution and a general shift in diet toward plant protein rather than on the need for birth control. See Frances Moore Lappé and Joseph Collins, with Cary Fowler, *Food First: Beyond the Myth of Scarcity* (New York: Ballantine, 1979). On the more general question of the human population explosion, see Frances Moore Lappé and Rachel Schuman, *Taking Population Seriously* (London: Earthscan Publications, 1989). Lappé and Schuman provide an excellent analysis of the power structures that contribute to high birth rates in developing countries. Although their primary focus is on the social causes and consequences of, and social solutions to, rapid population growth, Lappé and Schuman nonetheless argue (unlike most ecosocialists) that their analysis is capable of incorporating a nonanthropocentric perspective (see 70–71). In particular, they urge their readers to be cognizant of the impact of human population not only on humans but also on nonhuman life.

79. As Paul and Anne Ehrlich point out, the key to understanding the role of human population growth in the environmental crisis lies in the equation $I = PAT$ (with I representing environmental impact, P representing the absolute size of the human population, A representing affluence or level of resource consumption, and T representing the environmental disruptiveness of the technologies that provide the resources consumed). See Paul R. Ehrlich and Anne H. Ehrlich, *The Population Explosion* (New York: Simon and Schuster, 1990), 58–59. This basic formula was first published in P. R. Ehrlich and J. P. Holdren, "Impact of Pollution Growth," *Science, 171*, 1974, 1212–1217.

80. For a brief discussion of this holistic model as it relates to socialism, see Richard Worthington, "Socialism and Ecology," *New Political Science, 13*, 1984, 78.

INDEX

299

CONTRIBUTORS

Ted Benton is Professor of Sociology at the University of Essex, UK, and is the author of numerous publications including *Natural Relations: Ecology, Animal Rights and Social Justice* (1993), a member of the UK editorial board of *Capitalism, Nature, Socialism (CNS)*, of the editorial collective of *Radical Philosophy*, and of the UK's "Red–Green Study Group."

Robyn Eckersley is a leading writer on ecological politics and philosophy. She teaches politics in the Department of Politics at Monash University in Melbourne, Australia, and is the author of numerous publications, including *Environmentalism and Political Theory* (1992).

Hans Magnus Enzensberger is a prominent German poet, essayist, journalist, and dramatist. His *Dreamers of the Absolute* was published in 1988.

Arran Gare teaches in the Department of Philosophy and Cultural Inquiry at Swinburne University in Hawthorn, Victoria, Australia. He is the author of *Postmodernism and the Environmental Crisis* (1995) and *Environmental Destruction and the Metaphysics of Sustainability* (1996).

Michael A. Lebowitz is Professor of Economics at Simon Fraser University in British Columbia, Canada, and the author of *Beyond Capital: Marx's Political Economy of the Working Class* (1992).

Enrique Leff, of the United Nations Environment Program, Regional Office for Latin America and the Caribbean, is Professor of Political Ecology at the National University of Mexico and the author of *Green Production* (1995). He is a member of the editorial board of *CNS*.

Mary Mellor is a professor in the Sociology Department at the University of Northumbria at Newcastle, UK. She is the author of *Breaking the Boundaries: Towards a Feminist Green Socialism* (1992), and is a member of the UK editorial board of *CNS*.

James O'Connor is Professor Emeritus of Sociology and Economics at the University of California, Santa Cruz, and Editor-in-Chief of *CNS*.

Valentino Parlato is an Italian journalist, a founder of the communist newspaper *Il Manifesto*, and Coeditor of *Capitalismo Natura Socialismo*, the Italian edition of *CNS*.

Michael Perelman is Professor of Economics at California State University and the author of, most recently, *The End of Economics* (1996) and *The Pathology of the U.S. Economy: The Costs of a Low Wage System* (1993).

Giovanna Ricoveri is an economist, a retired official of the leftist trade union Confederazione Generale Italiana del Lavoro—CGIL, and Coeditor of *Capitalismo Natura Socialismo*, the Italian edition of *CNS*.

Gunnar Skirbekk is a Swedish geographer and the author of *Rationality and Modernity: Essays in Philosophical Pragmatics* (1993).

Kate Soper is a senior lecturer in philosophy at the University of North London, UK. She has written extensively on politics, philosophy, and feminist issues, and was a prominent activist in the Campaign for Nuclear Disarmament. She is the author of *Humanism and Anti-Humanism* (1986), *Troubled Pleasures* (1990), and *What Is Nature?* (1995).

Victor M. Toledo is an ecologist at the Centro de Ecologia, U.N.A.M., Mexico City, and a member of the editorial board of *CNS*.

Jean-Guy Vaillancourt is Professor of Socialogy at the University of Montreal, Canada.

Andriana Vlachou is Professor of Economics at the Athens College of Business and Economics, Greece, and a member of the editorial board of *CNS*.